D0961830

nerve

Adventures in the Science of Fear

EVA HOLLAND

NEW YORK

NERVE: *Adventures in the Science of Fear*

Originally published in Canada by Allen Lane, an imprint of Penguin Canada, a division of Penguin Random House Canada Ltd., in 2020. First published in the United States in revised form by The Experiment, LLC, in 2020.

The Experiment, LLC
220 East 23rd Street, Suite 600
New York, NY 10010-4658
theexperimentpublishing.com

This book contains the opinions and ideas of its author. It is intended to provide helpful and informative material on the subjects addressed in the book. It is sold with the understanding that the author and publisher are not engaged in rendering medical, health, or any other kind of personal professional services in the book. The author and publisher specifically disclaim all responsibility for any liability, loss, or risk—personal or otherwise—that is incurred as a consequence, directly or indirectly, of the use and application of any of the contents of this book.

The Experiment's books are available at special discounts when purchased in bulk for premiums and sales promotions as well as for fund-raising or educational use. For details, contact us at info@theexperimentpublishing.com.

Library of Congress Cataloging-in-Publication Data available upon request

ISBN 978-1-61519-600-5
Ebook ISBN 978-1-61519-601-2

Cover and text design by Beth Bugler
Cover photographs by Pascal Goetgheluck/Science Photo Library
Author photograph by GBP Creative

Manufactured in the United States of America

First printing April 2020
10 9 8 7 6 5 4 3 2 1

For my mom,
Katherine Janet Tait
(1954–2015)

I do not wish to fall short of your expectations in this matter. I intend, on the contrary, to treat the problem of the fear of nervous people with great accuracy and to discuss it with you at some length.

—Sigmund Freud

There are few things more liberating in this life than having your worst fear realized.

—Conan O'Brien

contents

prologue

We started the day with strong coffee and a short drive south on the Alaska Highway, from a remote lodge to an even emptier stretch of snowy pavement. As the dark February day grew lighter, we laced up heavy ice-climbing boots, buckled on packs loaded with ropes and gear and food and water, and began our ascent into the mountains.

It was February 2016, and our group of a dozen or so had come from our home in Whitehorse, the small capital city of the Yukon, a few hours away, for an extended weekend of ice climbing in far northern British Columbia. My friends Ryan and Carrie, and their crew of climbing pals, had been making this trip annually for several years. This was my first time tagging along.

Ryan and Carrie are natural teachers and leaders, people who genuinely enjoy passing on their skills and knowledge to others, and for the past few winters, they had been making occasional efforts to teach me how to ice climb: to ascend frozen waterfalls using crampons, axes, and rope. I was a poor student. I liked the *thunk* of my ax sinking solidly into thick ice, the soreness in my shoulders and calves as I moved up a route, step by step. I loved the glow of satisfaction when I reached the top of a climb. But I was afraid of heights—specifically, I was afraid of falling from exposed heights. Climbing, then, was hard for me. Ryan and Carrie had both seen me cry, more than once. They'd heard me beg to be allowed back down to the ground; they'd heard me announce, loud and flat and on the verge of losing control, that I was "not having fun anymore."

I had kept at it because, some of the time at least, I *was* having fun—and because I wanted to learn to master my fear. But my progress had been slow, and this winter I had hardly climbed at all. My mom had died suddenly the previous summer, and I had largely let sports and socializing slip away in the months afterward.

It was about an hour's uphill hike, alongside a creek bed through the snow, before we paused to strap on our crampons and then carried on up the frozen creek itself, the steel spikes offering us traction. The creek rose in slow increments: an easy step up, then a flat surface for a few steps, then a longer step up, and so on. Sometimes an icy rise would be too much to clear in a single step, and we'd kick the toe spikes of our crampons into its sloped surface and climb up that way.

Eventually we came to the true start of the climb, a route known as the Usual. One by one, we tied a rope to our harnesses and then ascended the first short wall. After that came another, longer, section of steep climbing, and then another.

It was a beautiful day, sunny and clear, with the temperature hovering around freezing. I was nervous, as usual—especially since there were members of the group that I barely knew. I was always even more mortified to show my fear in front of strangers.

But I was handling the climbing just fine, with no tears or pleas for mercy. I was even managing, as I sometimes did, to enjoy it.

When I made it to the top, I popped out onto an open, frozen plateau with a sweeping view all the way back to the highway. I took a selfie with the vista behind me and then sat down in the sun to eat my lunch, feeling proud and satisfied.

Around two o'clock, Carrie found me and suggested that I be one of the first to start making my way back down. I would likely be among the slowest. I agreed. Descending involved a series of rappels: tying into a secured rope and then lowering myself by hand down the ice walls.

I had never tried rappelling. The gang had taught me the basics the night before, in the hotel. They'd hitched me to a rope tied to a post in the hallway, and I'd walked backward across the linoleum, feeding the rope through my harness as I went. Of course, that lesson had taken place on a horizontal surface.

Still, I was feeling good, feeling ready. Carrie got me set up, and I steeled myself to walk backward off the edge of the plateau and down the icy face of my most recent climb.

The first rappel went well. I was able to laugh after I lost my balance, failed to brace myself with feet wide, and swung sideways into the ice. The second one was OK, too. The third was trickier: I had to make my way along a curving tunnel of ice, and again I lost my footing and swung hard into an ice wall, dangling helplessly on the rope, banging my elbows and knees. I swung and fell again as I neared the bottom of that rappel, and I slid down the rope and landed in a heap, the ice axes I'd hung on my harness digging hard into my sides.

Now I was embarrassed and in pain. I had a quick little cry there, tangled in my gear at the bottom of the rope, and then I picked myself up and moved out of the way so someone else could follow me down.

I struggled through two more short rappels, but my mood had become grim, and my control only deteriorated further as I descended. The afternoon was cooling off as sunset approached, and all that swinging and slamming had soaked me with water from the sun melting the ice through the day. I was cold, hungry, and exhausted. I was *not having fun anymore.* At the bottom of the last rappel, I sat down off to one side, away from the others, and cried, trying to hide my face. I ate a Snickers bar I'd been saving—chocolate almost never fails to cheer me up—but it only helped a little. We still had a long way to go to reach the cars.

When everyone had arrived at the bottom of the last rappel, we bunched together to head down the walkable, un-roped portion of the route, along the frozen creek. As the group began to head down in twos and threes, I stood on the edge of one of the low ice bulges I'd stepped up without difficulty that morning. There was maybe a foot, a foot and a half, between the flat surface I stood on and the next flat section of ice. All I had to do was reach out with my boot and step down. I stared at my feet, but I couldn't make them move. I kept picturing myself stepping down and my crampons failing to catch in the ice, my foot flying forward like I'd stepped on a banana peel in a cartoon. From there, I watched my body collapse, slide down the first ice bump and then the next, picking up speed, sliding and sliding down every frozen rise, all the way to the bottom. I couldn't do it, said a voice in my head. I would fall. I would die.

prologue

Some irrational force had taken over my body. I couldn't breathe properly, couldn't move my limbs. A tiny part of me knew I just needed to take one step down, that everything would be fine if I could only move my feet, but that voice of reason had been shoved into a corner at the back of my brain. Another voice was in control now.

Ryan noticed my distress and circled back to reassure me. I heard myself tell him that I couldn't come down the mountain, unfortunately. The group would just have to leave me there, I said. I couldn't walk down, so they would all have to go on. I would stay right where I was.

My flat tone said my plan was reasonable. But staying where I was, as the temperature plunged and darkness came on and I stood shivering in my wet Gore-Tex, would be suicide. Still, my feet refused to move. I watched Ryan confer with the others and send Carrie and the rest of the group on ahead, so they could make it to the cars before dark. Ryan, his friend Joel, and a third guy I barely knew, Nic, stayed behind.

Joel stood on one side of me and grabbed my left hand. Nic took hold of my right. Ryan stepped down to the next ledge and turned to face me, pointing with his ice axe at the spot where I needed to put my foot. Slowly, taking deep breaths and clutching Joel's and Nic's hands hard, I forced my right foot down. My crampons caught. I did not slide to my death. Then we repeated the process with my left foot.

The light dimmed, and the night got cold. We inched down the mountain, Ryan pointing out every step, promising me that it was safe. Right foot. Left foot. I think I cried quietly, some of the time, from fear and frustration at the extent to which my body and mind had betrayed me. I was still halfway convinced that if I took one wrong step, it would be the step that killed me. It felt like the descent took hours. Eventually we pulled out our headlamps and carried on downhill in the dark.

Once we made it off the ice and onto the snowy trail for the last part of the hike down, I was finally able to let go of Joel's and Nic's hands. We tramped to the highway mostly in silence, and my fear receded enough for me to wonder how angry they were. Did Ryan wish he'd never invited me on this trip? Surely he must. By the time we were in the car, the four of us piled into the single remaining vehicle, my lingering fear had been eclipsed by the most powerful feeling of humiliation I have ever experienced. I sat in the back seat, trying to shrink into nothing, unable even to enjoy Ryan's traditional post-climb bag of dill-pickle-flavored potato chips. I was utterly mortified.

Back at the hotel, I did the best I could. I forced myself to socialize with the group over cards and drinks instead of hiding myself away. I offered beers to the guys who'd peeled me off the mountain. At some point, I asked Ryan what he would have done if I hadn't voluntarily taken that first downward step. "You wouldn't have liked it," he said. I had a vision of Ryan and Joel hog-tying me and dragging me down the frozen creek, bump by bump, in a slow-motion version of my imagined death-by-sliding. He was right. I wouldn't have liked it.

The next day, when the rest of the group went climbing again, I stayed at the hotel. I went for a long run along the highway. I read a book. I tried to relax and enjoy my weekend, tried to appreciate the blue sky and the white mountains all around the lodge. But I kept thinking back to my behavior the day before.

It was unacceptable, I decided. I'd tried half-heartedly to work on my fear of heights over the years, but the matter had never seemed urgent. I had never before put my own life, and the safety of others, in danger because of it. I could hardly believe the lunatic on the mountain had been *me*, declaring that I would die from exposure rather than walk down a frozen creek. What was the matter with me?

I tried not to let myself dwell on it, but my collapse on the Usual was a setback when I had only just begun to put myself together again. For much of my life, I had feared my mom's death. Her own mother had died when she was a child, and growing up, I had become intimately aware of the devastation the loss had left in its wake. I had dreaded living through the same loss, and when my turn did come, I had fallen apart. In the months after, I had retreated completely from my life: from friends, from exercise, from the things I normally did to challenge and amuse myself. For too long, I felt like I had forgotten how to smile or laugh, like the muscles in my face had stiffened up and no longer knew how to perform those simple acts.

It had been only a few weeks since I'd started to reengage socially. I had started running, started feeding myself properly again, stopped living on my couch in a blur of binge-watched TV shows.

I didn't want my setback on the mountain to derail my slow, hard-earned return to normal life. I didn't want my terror to control me that way ever again. I decided, sitting alone in that hotel by the side of a lonely highway, that I would figure out what had happened in my brain on the mountain that day. And then, I decided, I would figure out how to fix it.

Over the next few weeks and months, I began what I would think of as my fear project. I checked out books from the library, clearing the self-help section of its various inspirational face-your-fears titles, and picking up everything I could find about the science of fear and phobias. I started talking to my friends and family about their own fears—everyone, it seemed, had a story to tell about the things that terrified them, and theories about why, and how, and hearing those stories helped expand my understanding of just how big a role fear can play in all our lives. Most importantly, I started plotting ways to try to conquer, or overcome—or at the very least renegotiate my relationship with—my fears.

I came up with three broad categories of fear to pursue. The groupings were imperfect, sometimes bleeding into one another and by necessity not encompassing every flavor of fear, but I figured they were a starting point for understanding.

First, and most obvious, are phobias: clinical, seemingly irrational fears, mostly connected to elements in the world outside ourselves. In my case, they are represented by a potent but narrowly focused fear of exposed heights. Then there is trauma, phobia's more concrete cousin—the fear that lingers in our bodies and minds after bad things happen, driven at least as much by our fear-filled memories as by a fear of our possible future. Trauma is most closely associated in the popular imagination with personal exposure to violence, but for me, trauma was a legacy from a series of car accidents.

Lastly, there is the ephemeral, hardest-to-pin-down suite of existential fears that seem to come as part of a package deal with our human consciousness: our fear of dying, our fears of loss, our uncertainties about the world and our place in it. This tangled mess of fears is most prominently represented in my own life by my fear of my mother's death—and it was her death, my worst fear come true, that sent me off in search of a greater understanding of all my fears, and the fears we all carry.

Understanding the ways that fear had threaded its way through my own life also meant learning about how fear acts and reacts in our bodies and minds. It meant trying to trace the connections between phobias, anxiety, and trauma, and the ways our society has responded to each one over time. It meant examining all the times in my life when I've felt truly, deeply afraid, dusting off old memories and trying to parse them. Had my fear been rational, a justifiable response to a threat? Or was it overblown and even toxic, like my reaction on the Usual? If fear was an essential survival tool, why did mine sometimes seem to lead me into greater danger?

Every answer I found brought new questions. And as I threw myself into various treatments for fear, both medically sanctioned and homegrown, every solution I tried brought new insights—even if they didn't always bring me relief.

This book is the end result of all those questions and answers. After everything, I can't say that I am now in perfect control over my fears. I can't even promise that my breakdown on the Usual will never occur again. But I can say that my relationship with fear will never be the same.

PART ONE

1

a personal history of fear

My worst fear came true on a summer evening in July 2015. I was sitting on a log by a foggy river, in the northwesternmost corner of northwestern British Columbia, when some number of weakened blood vessels lurking in my mother's brain let their walls down. It was a Friday night. I was eating red curry with tofu, drinking beer, singing camping songs, on a ten-day backcountry canoe trip with eleven friends. Three time zones away, in an Ottawa restaurant, my mom was out for dinner with my stepdad and another couple. They'd just paid the bill when she said, "I have the most terrible headache." And then she said, "I think I'm having a stroke." And then she collapsed, and an ambulance was called, and the paramedics arrived and knew on sight that it was bad, really bad.

I was probably singing "Wagon Wheel" or "Cecilia" when they got her to the hospital, to the intensive care unit, where they riddled her with needles and put a tube down her throat, drilled a hole into her skull and stuck a tube in there, too.

For the rest of that night and for the next three days, a machine helped her breathe. Monitors beeped. Nurses bathed her and turned her. I paddled my canoe to a natural hot spring, soaked in its murky waters, smelled the wild mint that grew all around. I made camp and broke camp, got sunburned, picked blueberries. Meanwhile, my family tried to find me. They left messages on my land line, on my cell phone, on Facebook. But I was a long way from the nearest communications tower.

A cousin, Nathan, was my emergency contact for the trip. He was a biologist who worked in fisheries management—and luckily the salmon were running. By Monday morning, he was able to reach a Canadian government fisheries outpost that was set up along the Stikine, the river I was paddling. He had our itinerary, and he knew that we were due to spend Sunday and Monday nights at a campsite in Great Glacier Provincial Park, the only formal, developed campsite we would visit on the trip, and the only time we would spend two nights in the same place. On Monday, while we portaged our canoes from our riverfront campsite along a narrow, densely treed trail to a glacial lake, while we paddled across the lake to the steep face of the Great Glacier and craned our necks to stare up at its blue-and-white corduroy folds, a fisheries officer motored upriver and found our camp. On a picnic table, under a tarp we'd strung up, he left a note pinned down by a rock: EVA HOLLAND YOU MUST CALL NATHAN IMMEDIATELY REGARDING YOUR MOTHER.

I was among the last ones back to camp. Carrie and I were carrying her canoe on our shoulders, me in the front, only my shoes and the path visible, when I heard Ryan up ahead. In a weirdly calm, flat voice, he asked us to put down the canoe. Then he handed me the waterproof case that held our satellite phone, brought along for emergency use only, and told me I had to call Nathan. I was puzzled at first, my rising concern slowly washing away the residual high from running a rapid with Carrie to save a few hundred yards on our portage. But when I saw the note, the world froze. I'd been afraid of this moment for a long time, and now, I knew, it had found me.

Ryan led me to a quiet area of the campground and showed me how to dial the satellite phone. I clutched his hand hard while it rang, and then Nathan picked up. I could hear the strain in his voice, his impossible search for the right words. My mother had had a stroke, he said. He struggled to say more. I heard myself ask, in a voice so neutral it was almost cold, "Is she still alive?" She was, he said, but she was unconscious, and no one was sure if she would ever wake up. Tests were being done. There was a waiting period to be endured. "I think you should get there," Nathan said. I almost laughed. How was I supposed to do that? But he had a plan.

Logistics were beyond me, so I passed the phone to Ryan, who had a notebook and a pen. Ryan spoke tersely to Nathan while I collapsed to the ground and sobbed into the spruce needles.

The plan went like this: I would remain with my friends on the river for one more night, and tomorrow we would set out as we had originally planned. There was a salmon processing plant a couple of miles downstream, and they were making a delivery run to Wrangell, the Alaska town near where the Stikine met the coast. The workers would take me along for the ride, covering in a matter of an hour or so a distance we would have spent three more days canoeing. From Wrangell, I would fly to Seattle and beyond. Nathan would book the flights. If all went as planned, I would be in Ottawa in thirty-six hours.

It was a rainy, grim evening. I joined the group for supper and tried my best to act like a normal, healthy human being—like a person on a canoe trip with her friends rather than someone waiting to travel across the continent to have her worst fears confirmed. We talked a little about what was happening; a couple of others in the group had recently lost parents. One of them told me it was important that I get there in time—before she finished dying, he meant. I felt safer, somehow, knowing that there were people in my life who had been down this path before me. There was wisdom to be gleaned from others about how, exactly, I was supposed to proceed.

That night, I kept my tent-mate awake with my crying, although I tried to be as quiet as I could. I didn't sleep much, and the next morning I felt stretched thin, exhausted. I was afraid to be leaving my friends and facing the journey to Ottawa alone, and I was afraid of what was waiting for me there. We floated easily downriver to the plant, where the staff gave us a tour and let me use a shower before my friends shoved off downstream again.

We passed them soon after in the laden boat and waved another set of goodbyes. Then I was really on my own. I rode out to the coast sitting cross-legged on a stack of large plastic tubs loaded with seven thousand pounds of fresh Chinook salmon.

On the dock in Wrangell, I sat in the sun and called my dad on my cell phone while the workers ran errands. The rest of my family had initially been unable to reach him, but he was up to speed now, and I was relieved to hear his voice. I don't remember what we talked about for those fifteen or twenty minutes, except that at one point—probably amazed by the looming void that was about to open up in my life—I mentioned that my mom and I usually talked on the phone four or five times a week. There was a pause on the line, and in a moment of sudden, clear insight, I could almost hear him struggling not to show his shock, to bury any hurt feelings for the sake of handling the larger crisis. He and I talked once a month or so, maybe a little more, and though he'd always known I was very close with my mother, I don't think he'd ever had the difference confront him so bluntly before.

"Well, you can always call me," he finally said. I agreed.

We were already planning for a future without her. During those thirty-six hours, from my satellite phone call to my race across the continent, I don't remember anyone suggesting to me that my mother might get better, that she might wake up.

I was late to the airport, too late to check the waterproof bag with the handful of items I'd carried with me off the river. The bag was too big to carry on. Even this one small obstacle was too much for me, and I broke down crying in the check-in line, blurting out between gasping sobs, "My mom is dying." The airport staff confiscated my toothpaste, my sunscreen, all my liquids, and let me board with the bulky bag in hand. I wondered later if they'd thought I was lying, that I was some kind of nut job with a penchant for drama. I wondered if they heard that kind of thing every day. I spent that first flight to Seattle staring out the window so my seatmate couldn't see me crying. *My mom is dying*, I thought to myself, over and over again. *I am never going to talk to her again.* I was trying to make this most-feared moment real, trying to harden myself against the reality that I was flying toward at five hundred miles an hour.

I ran through the Seattle airport to catch a red-eye to Toronto, where I cleared customs and then boarded a short-hop commuter flight for the last leg. My dad met me at the Ottawa airport, waiting at the bottom of the escalator, as he always did. I went to his townhouse long enough to shower, and then we drove to the hospital, me wearing the Gore-Tex rain pants and Crocs that were the only cleanish clothes I'd had with me on the river. My mom's sisters, my aunts Shelagh and Rosemary, were already there. Tom, my stepdad, had been sleeping on a couch in one of the quiet family consultation rooms in the ICU since Friday. It was now Wednesday morning, and there was no end in sight. Everyone shuttled between the ICU waiting room, the coffee shop downstairs, and the private room where my mother was lying among her tubes and wires, her chest rising and falling in time with the whirring ventilation machine.

It was clear that I was expected to spend time sitting with her. The expectation seemed to be that I might even enjoy that time, on some level. But seeing her there appalled me. She would have hated it, I thought—hated us all seeing her that way. Her face was puffy and pale, her hair was dirty, and further sullied by the rusty red antiseptic the staff must have splashed around while they were affixing all her wires and tubes. My mother was fastidious about her appearance; I had tried several times over the years to get her to come to New York City with me, a place I loved and visited regularly, and she had always refused, claiming that she didn't have the wardrobe for it.

A kind nurse settled me down in a chair next to the bed and retrieved my mother's hand from under the blankets for me to hold. Wires and tubes intruded, but I reached out hesitantly and touched the soft skin on the back of her hand. That much, at least, was familiar. That much I recognized.

The attending physician in charge of my mother's care was tall and broad-shouldered, with dark hair and dark eyes. My aunts and I nicknamed him Dr. Handsome. Over that day and the ones that followed, he was the one who would gather us for updates in a private meeting room, speaking carefully and solemnly. I had thought, during my race across the country, that the decision had already been made—that my family was just waiting for me to arrive before they turned off the machines. But it turned out things weren't yet so certain.

That afternoon, my mother's test results came back, and Dr. Handsome let us know that the news was not good. The stroke had occurred in her brain stem, where so many of our vital life-giving functions are controlled, and there was no possibility of repair or recovery. She would never wake up; she no longer had the capacity to do so. She would never breathe on her own again. She was no longer aware of her own existence.

We could wait, keep her on the machines and see if, somehow, she improved. She hadn't left any guidelines about end-of-life care. She was only sixty and had been physically healthy up to now. Dr. Handsome and his team didn't push, but they let us know that the most likely outcome of waiting was infection. Still, it was up to us to make the call.

Over the next forty-eight hours, I spent most of my time at the hospital—in the ICU waiting room with my aunts and friends from high school who rotated through when they could get away from work and child care, or in the meeting room with Dr. Handsome and everyone else, weighing our options, or alone in my mom's room again, holding the hand that the nurse had fished out from under the covers. "You can talk to her if you'd like," the nurse had said. I tried, but it felt strange. She wasn't there anymore—Dr. Handsome had said so. Any talking I might do, I supposed, was for my own benefit, not hers.

On Friday afternoon, the decision was made. We would discontinue life support—after which, it was assumed, my mother's body would die within minutes, maybe an hour.

I had asked my dad to stay with me for the event itself. We all gathered in the small hospital room: my stepdad Tom, me, Shelagh and my uncle Peter, Rosemary, my dad, two close friends of Tom's, Dr. Handsome, and the other relevant ICU staff. My cousin Bobby walked in as the procedure was beginning, not realizing when he'd begun the drive to the hospital that normal visiting had come to an end. He hovered in the doorway, unexpectedly a part of the death vigil, unsure if he was welcome but also reluctant to turn and walk away.

I sat on my mom's right side and held her hand. Across from me, Tom did the same. My dad stood behind me, a hand on my shoulder; I reached up and clutched it. Moments earlier, the nurses had ushered us out briefly while they removed the tubes and wires—so my mom would look a little more like herself in the final minutes, I suppose. All the machines had whirred to a halt.

Dr. Handsome had explained to us already that her ability to breathe on her own was impaired but not completely destroyed. That turned out to mean that we would witness not a peaceful passing away but a desperate final struggle for oxygen. As the minutes ticked down, she gasped like a fish in the bottom of a boat; she made small, terrible noises. I had been led to believe that my being there was important, that I would be glad, someday, to have seen the end—that I would be grateful. I did not feel grateful.

"Jesus Christ," I said, grinding the words through clenched teeth, after one sharp gasp, and Dr. Handsome reminded me again, in his calm, solemn voice, that my mom had no awareness of her own existence anymore. She couldn't feel pain. She couldn't know that she was dying, he promised—that we were allowing her to die.

It took about twenty minutes. The color drained from her face, her lips turned grey, and the hand in my hand began to cool. Somebody made the official pronouncement. We sat for a few minutes longer, and then we got up and filed out.

My dad drove me home to his townhouse. I stared through the windshield, struggling to fathom a future without my mom. I should have been shocked by my loss, I suppose; it had been so sudden, so fast. "You sit down to dinner, and life as you know it ends," Joan Didion wrote in *The Year of Magical Thinking*. And that was true. When I'd walked out of the hospital and gotten into the car, it had felt like the first few minutes of a strange new life: I was still Eva, but I was different now, fundamentally changed.

Still, while I was unmoored, I wasn't surprised. On some level, the fact that I was now, abruptly, motherless made perfect sense to me. It felt inevitable. I had been bracing for my mom's death, fearing it, for almost as long as I could remember.

PEOPLE OFTEN TALK about a fear of the unknown, and fair enough—we do tend to be wary of the strange and new. But we can also learn to fear what we know.

I grew up knowing that my mom was an orphan. Her mother, Janet, had died of colon cancer when Janet was forty-five and my mother was ten, and her father, Robert, had followed when my mom was nineteen. I don't remember anyone explaining these sad facts to me; they were just woven into the fabric of our lives, our family story. My dad's parents, the people I called Grandma and Grandpa, were the grandparents I knew. But the two ghostly strangers on my mother's side were always a part of my life, too.

Janet had been beautiful and vivacious, I knew. In my mom's faded childhood memories, she was the kind of woman who walked down the street singing out loud, not caring who might hear. In the handful of small black-and-white photos my mom kept in an old cigar box, Janet had big dark eyes, stylish 1960s hair, prominent cheekbones, and a wide smile. Before she was married, during the war, she had traveled across Europe with some of her best friends, singing for the troops; some contested family lore held that she had once been on a date with a then-still-single Prince Philip. In photographs, she was always smiling, her teeth strong and white, the arch of her brows and the curve of her lips demanding your attention. Bob, an air force war veteran and metallurgical engineer, was a less memorable presence, neatly groomed and bespectacled and always seeming to fade into the background of the photographs, content to be eclipsed by the bright star of his wife.

My mom was the youngest of their three daughters, and her early years were happy. She was the bright, charismatic baby of the family. (Confidence, even approaching arrogance—the faint smugness of the youngest child—is visible in her childhood photos.) Her mother was beautiful and well liked; her father was serious and successful. The family moved several times for her father's career, making a tour of various Canadian mining towns, and by the time Janet got sick, they were living in Lakefield, Ontario, in the quieter country northeast of Toronto. Things got murky then. In my mother's recollection, the cancer was a near secret, handled almost stealthily, treated as a subject for grown-ups when it was spoken of at all. She didn't remember being allowed to say goodbye to her mother. She was pretty sure she wasn't even brought along to the funeral. She never knew where her father placed Janet's remains. One day, she had a furtively ailing mother; the next, she had none.

Bob unraveled after his wife's death. He withdrew from his daughters' lives and turned to alcohol. He had the girls shipped off to various boarding schools and distant relatives' homes. He remarried hurriedly, to a woman with five children of her own.

My mom, as the youngest, was the only one of the sisters to spend significant time in the house her father now shared with this new family. She had fond memories of her step-siblings, but she remembered their mother as cruel, resentful of the broken girl in her care. In my own childhood, I understood the archetype of the wicked stepmother, and I assimilated my mom's story as a kind of fairy tale. Here, in my child's mind, was a real-life Cinderella, waiting for deliverance, toiling away while her father became indifferent in his grief.

She shuffled through four different high schools at opposite ends of the country. She started to attend university, but then Bob died three days after her nineteenth birthday. Now orphaned, she dropped out of school, moved in with her middle sister, Rosemary, in Toronto, got a waitressing job, and immersed herself in a hippie scene centered around the food co-op movement, where she eventually met my dad. She was twenty-four when she married him and twenty-seven when I was born.

That apparent happy ending didn't mean she was OK. When I was still very young, I became aware that the loss of Janet, in particular, had left a lingering injury in my mom. I knew that she got sad sometimes, that on some days she just needed to stay in bed and be left alone. In the year after my parents split up, when I was seven, those "some days" became almost every day. And in that first year after the divorce, while the two of us rattled around a big three-story rental house together, I began to understand that Janet's death was the central fact of my mom's life.

One night, during that long year, my mom and I had a terrible fight. I don't remember what I was so enraged about, what started things off, but I remember how it ended: I ran up to my bedroom, grabbed a small framed photo of Janet that I kept on my bedside table, and held it over the banister, threatening to smash it. My mom came undone, sobbing and begging me not to do it, and I was so terrified by the sudden reversal of the authority between us that I immediately put the photo away. I had glimpsed for the first time that I had immense power over my mother—that despite my status as child, and hers as parent, she was vulnerable. Her mother's death had left her that way.

Seeing her sadness up close, without another adult there to intervene between us, made me understand that our dependence on one another, emotionally at least, was mutual, even if she was still the one who provided the grocery money. When I raged, she cried. When I fought dirty, attacking her weak spots, as perceptive children soon learn to do, I left a mark. I began to realize then: No one can hurt us more deeply than the people we love, and that's part of why love and fear are so tightly bound up together. We want to protect our loved ones; we fear being the ones to hurt them, maybe as much as we fear being hurt by them in turn.

After that, I never viewed my mom entirely as an authority figure. She would be a roommate, a guardian, a confidante, a best friend, a trusted expert on any number of topics, the first person I wanted to talk to about anything of importance. But I was governed after that year not by any fear of punishment, of the consequences of disobedience, but by the fear of upsetting her. My guiding principle—which far too often I failed to uphold—was to avoid making my mother cry.

We moved cross-country, to Ottawa, following my dad's job, and fell into a routine. My mom finished her degree, got a job doing administrative work for a women's advocacy group, and I went to school, alternating weeks between her place and my dad's. On her end, Saturday nights were for old movies on the local public television station. Special occasions meant dinner at the

Indian restaurant around the corner. I loved our lives together, but I was always aware of her sadness. It felt like a warning of my own future pain.

Still, even on her worst days, my mom always had time to talk, to ask me about my life or hear my concerns and offer advice. (If that meant we talked while she lay in bed, then that was how we talked.) She was smart and funny, and unconditionally supportive. She used to greet me after school by asking, "What did you learn today?" That always made me roll my eyes. If I didn't want to talk about school, she'd follow up with "Tell me your hopes and dreams, then." And that made me roll my eyes harder.

I wasn't the kind of kid who signed up for a full slate of after-school activities, and we spent most of our evenings together, watching *Dallas* and then *Street Legal* and then *NYPD Blue* as the years passed. We watched the Blue Jays win the 1993 World Series from her queen-size bed, our little TV set rolled into the room on a cheap Ikea cart. We watched the Olympics together every time they came around, re-familiarizing ourselves with the decathlon, the butterfly, the salchow, and the lutz. Sometimes we were rebellious coconspirators in our little apartment: If we didn't feel like cooking, we ordered pizza for dinner, or just guzzled bowls of Honey Nut Cheerios.

In 1994, when I was twelve and my mom was approaching forty, a writer named Hope Edelman published *Motherless Daughters: The Legacy of Loss.* The book was partly a memoir of Edelman's own mother's premature death and partly a compendium of research and collected anecdotes about the impact of a mother's death on young women at different stages of their lives. If I had picked it up and read it back then, I would have found a remarkably accurate portrait of my mom and her pain and her fear that, having lost her own mother, she would never be an adequate mother or woman herself.

My mom bought the book in hardcover soon after its release, and though I don't know for sure if she ever read it, it took up permanent residence on the little wicker bookshelf next to her bed. I would notice it there nearly every time I came in the room, stacked with other self-helpy titles and gathering dust, and as the years passed, it became a symbol to me of my mom's ongoing loss. That was how she defined herself, I understood by now: forever a daughter, forever lacking that most important figure in her life, her mother.

Lying there on my mother's bookshelf, staring back at me, the book also felt like an omen of my own possible future. I didn't want to be sad and hurt, like my mom. Above all else, I feared becoming a motherless daughter, too.

That wasn't my only fear when I was growing up, of course. Like most people, I was prey to an array of things that scared me. (An irony: Fear is an experience that unites us, even as, in the moment, it makes each of us feel alone.)

In my earliest clear memory of being afraid, I am three, maybe four, years old. I'm standing at the top of a long, descending escalator in Toronto's Pearson International Airport, on my way to my grandparents' house in the distant suburbs. My parents are with me, my mom and my dad. Probably one of them is holding my hand. I don't remember for sure.

What I do remember is putting one foot on the escalator's moving top step and then, suddenly, being struck with a fear of falling. So I do what people do when we're afraid: I freeze, with one foot on the escalator and my other foot still on the solid airport floor. Inevitably, the escalator churns on, my little legs split apart, and I tumble down a couple of steps, the serrated metal edges of each one scraping long red claw marks into my shins. Imagined consequences had produced real ones; my fear had been a self-fulfilling prophecy.

That fear of falling, the one that hit me at the top of the escalator, resurfaced at other places and times. When I was eight or nine years old, I came home from school one day and confessed to my mom that when we ran races in gym class, I never went as fast as I could go. I held back, I said, because I was afraid of falling, of losing control. She always loved that anecdote. She thought it said a lot about my character, my essentially cautious nature. When she retold it over the years, though, I never really liked the way the story made me feel. Who wanted to lose because they were afraid to take a chance?

I remained afraid of descending escalators for years after my tumble at Pearson. Into high school, I had to free both hands, count to three, and grip the handrails before I could make myself plant both feet on that top step. Even today, I take a deep breath before I step onto one.

Those were manageable twinges, though. I was fifteen years old when I had my first real attack of irrational, off-the-charts fear. It was the summer after ninth grade, and I'd signed up to spend a week on an old-fashioned tall ship on Lake Ontario with a dozen other teens, a sort of summer camp under sail.

I loved everything about life on board that ship: sleeping in my narrow metal bunk below deck; waking in the middle of the night to stand watch, peering out at the endless darkness; lounging on sunny afternoons in the thick rope net that hung below the ship's carved wooden bow. On deck, we wore basic harnesses around our chests, fitted with a short rope ending in a heavy metal clip. In very rough weather, or if we were climbing the mast to adjust the sails, we were meant to clip ourselves in, just in case.

The problem came the first time I tried to climb the mast—to "go aloft," in sailing terminology. I got partway up, moving my clip to each new rung as I went, fighting panic with every step up the ladder. My chest tightened and my breath shortened. My brain felt squeezed by fear. My muscles didn't want to obey me—each move was like pushing through wet cement. Then, finally, halfway up, I froze. I couldn't stop staring at the wooden deck, swaying below me, couldn't stop picturing my body splattering against it, my bones shattering, my blood running into the lake.

I couldn't make myself go forward, and I couldn't seem to retreat, either. The ship's "officers"—our camp counselors—eventually managed to coax me down, sending up soothing words and encouragement from below, and once my feet touched the deck, I never went aloft again. Everyone was kind to me about my failure, but there was no point in coming back the following year. A sailor who can't adjust the sails in a pinch isn't much use.

Even once I'd abandoned my short-lived sailing career, high school remained a minefield of things to be afraid of. I was afraid of doing hard drugs; I was afraid of *not* doing hard drugs if they were offered to me. I was afraid of the loud men who slowed their cars and yelled crude invitations, idling alongside me as I walked down dark streets at night; I was even more afraid of the ones who hovered silently.

Some of my fears were darkly specific: I was afraid of drinking too much and drowning in my own vomit. (At school we had been warned about this hazard repeatedly.) I was afraid of not having enough money for university, afraid of going away for university, afraid of staying home for university, afraid of making the wrong choice of university. (If we made the wrong choice, one recruitment rep who visited our high school assured us, we would end up working at McDonald's.)

As it opened itself up to me, even as I charged forward to meet it, the world seemed suddenly full of dangers—more than I ever could have imagined when my biggest concern was about running too fast on the playground.

Underlying it all, I carried a tangled thread of fears about my mom. I was afraid that I would hurt her, afraid that I would lose her, and afraid that, in losing her, I would become like her. I loved her, I admired her, but I didn't want to carry that same sadness in my own life.

During that year we'd spent alone, the two of us, in the big old rented house, I had come to understand three things with total clarity. First, that the loss of a mother could be a life-destroying force—as the loss of Janet had been for my mom. Second, that the same thing could happen to me if my mother died. And third, that because of my mother's immense sadness, she was vulnerable. It was up to me not to hurt her any more than she was already hurting, not if I could help it anyway.

So that was me as I reached adulthood: loved, cared for, mostly healthy, and carrying, as most of us do, a quiver of assorted fears. At the dawn of my twenties, I was still nervous on the down escalator. I was afraid of making the wrong choices, the wrong moves as a young adult—the wrong school, the wrong career, the wrong relationship. But if you'd asked me then to name my worst fear, I would have told you—if I was feeling honest—that there was nothing I feared more than my mother's death and the emotional destruction I assumed it would rain down on my life.

Fear wasn't yet something I thought about as a phenomenon in itself. I hadn't started to wonder about how it functioned in my body's cells, or even to think about whether my fears were something I could conquer, overcome, instead of sidestepping them or acquiescing to their power. I hadn't yet heard of post-traumatic stress disorder (PTSD), or given much thought to the existence of phobias. I hadn't yet asked myself what it would be like to live without fear, or to think about why fear is necessary, an essential component of our emotional lives, even as it feels like a hindrance or an embarrassment.

All of that's changed now. Now, I think about fear, and the questions that surround it, all the time. I guess being forced to face your worst fear, and come through the other side, can have that effect.

2

this is your brain on fear

Fear, it seems at first, should be easy to identify and define. To borrow from that old judicial decision about the definition of obscenity: We know it when we feel it.

Putting that feeling into words can be harder. G. Stanley Hall, the nineteenth-century founder of the *American Journal of Psychology* and the first president of the American Psychological Association, described fear as "the anticipation of pain," and that seems like a pretty good general definition to me. Fear of violence? Anticipatory pain. Fear of a breakup, the loss of someone you love? Anticipatory pain. Fear of sharks, of plane crashes, of falling off a cliff? Check, check, and check.

But what we need, really, isn't just a solid catch-all definition. What we need, to understand the role of fear in our lives, is to examine the layers and varieties of fears that can afflict us.

There's the sharp jab of alarm when you sense a clear, imminent threat: *That car is going to hit me*. There's the duller, more dispersed foreboding, the feeling of malaise whose source you can't quite pinpoint: *Something is wrong here. I don't feel safe.* There are spiraling, sprawling existential fears: *I am going to flunk this exam, tank this interview, fail at life.* And there are precise, even banal, ones: *Pulling this Band-Aid off is going to hurt.* How do they all fit together? Or, put differently, to what extent does each stand apart?

According to Greek mythology, Ares, the god of war, had two sons, who accompanied him into battle: Phobos, the god of fear, and Deimos, the god of dread. That seems like a useful distinction to start with—fear versus dread—and it's one that's echoed today by our distinction between fear and anxiety. Fear, generally speaking, is regarded as being prompted by a clear and present threat: You sense danger and you feel afraid. Anxiety, on the other hand, is born from less tangible concerns: It can feel like fear but without a clear cause. Simple enough, at least in theory.

In *Fear: A Cultural History*, author Joanna Bourke gamely attempts to parse the distinctions between fear and anxiety. "In one case a frightening person or dangerous object can be identified: the flames searing patterns on the ceiling, the hydrogen bomb, the terrorist," she writes. Whereas "more often, anxiety overwhelms us from some source 'within': there is an irrational panic about venturing outside, a dread of failure, a premonition of doom. . . . Anxiety is described as a more generalized state, while fear is more specific and immediate. The 'danger object' seems to be in front of us in fear states, while in anxiety states the individual is not consciously aware of what endangers him or her."

But as Bourke points out, that distinction has serious limitations. It's entirely dependent on the ability of the fearful person to identify the threat. Is it legitimately, immediately dangerous? Or is the fear abstract, "irrational"? She offers the hydrogen bomb and the terrorist as examples of potentially clear and present threats, but both can also serve as anxiety-inducing specters, ominous even when absent.

Consider my meltdown on that ice-climbing trip. I was convinced, utterly convinced, that the frozen creek represented a legitimate and potentially lethal threat to my safety. And sure, one *could* theoretically slide down a frozen creek to one's death, right? An ice-covered mountain isn't the world's safest environment, and that's objective fact. But in context, on that sunny but cold afternoon, was my fear—my fierce conviction, my paralyzed reaction, my refusal to move—a reasonable reply to an immediate threat? It very clearly was not.

The distinction between fear and anxiety, then, can be murky, even as it can also be a useful and even necessary line to draw. Many of the fear episodes discussed in this book might also be described as containing at least some elements of anxiety.

Then, setting the issue of a threat's clear presence aside, there's the matter of our "fear" response.

The scientists who study our emotional lives make distinctions between different categories of feelings. There are the primary emotions, our most basic and near-universal responses, found across cultures and even appearing, or at least seeming to us to appear, in other species: fear, anger, disgust, surprise, sadness, and happiness. Think of them like primary colors, the foundational elements of a whole rainbow of emotion. Just as red and blue in combination can be used to create all the shades of purple, you can imagine some more precise feelings as being built by the primary emotions. Horror, for instance, is fear mixed with disgust—and, maybe, some shadings of anger and surprise. Delight could be happiness with a bit of surprise stirred in. And so on.

There are also the social emotions, the feelings that don't stand alone like the primary emotions but are generated by our relationships to others: sympathy, embarrassment, shame, guilt, pride, jealousy, envy, gratitude, admiration, contempt, and more.

Of all these, fear is perhaps the most studied. But what does it really mean to study fear? What do we even mean, exactly, when we say "fear" in the context of scientific research? That's a more complicated question than you might expect.

Traditionally, scientists have studied "fear" in animals by measuring their reactions to threatening or unpleasant stimuli—a rat's freezing response when it is subjected to a small electric shock, for instance. In studying humans, scientists have more options and a broader array of tools. Most importantly, humans can self-report, verbally or in writing: *Yes, I felt afraid.*

The complicating factor is that those two responses—the freezing and the feeling—are separate and distinct. As the neuroscientist Joseph LeDoux, an expert on the brain circuitry of fear, emphasizes in his book *Anxious*, we know that the physical fear response and the emotional feeling of fear are produced by two different mechanisms in the body.

My interest is in both the physical fear response *and* the feeling.

One night when I was eleven or twelve years old, I had a terrible nightmare. The dream, as I remember it, was like a film shot in grainy black and white: I was in the ground-floor apartment that I shared with my mom, and, somehow, I knew that we were not alone. There were intruders in the house, though I couldn't see them, only the apartment's dim, grey hallways, the candelabras silhouetted on the walls. I knew that the intruders meant to kill us.

I woke up in my bed with the dream unfinished, my mind still filled with dread. I got up, walked across the hall to my mom's room to climb in with her, and fell back to sleep.

A couple of hours later, I was ripped awake again by a tearing, stabbing, burning pain in my left knee. I woke up screaming, convinced that someone was putting my leg through a meat grinder. I only had a moment to understand that my leg was fine, that the pain was in my mind, and to register my mom's fear and confusion, before I began to convulse. My legs and arms thrashed wildly, my spine arced and released, arced and released. It felt weirdly rhythmic; my body was throwing a party that my mind had not been invited to. Somehow I wound up on my stomach, and I can still so clearly remember the feeling of my neck jerking my head back and then forward again, over and over. I kept trying to scream but always wound up with a mouthful of pillow.

I faded in and out of consciousness for that first seizure, but at some point—after thirty seconds, maybe, or a minute at most, though it felt far longer—I became aware that the convulsions had stopped.

I seized twice more before I received a formal diagnosis of epilepsy. The next two came in the same night, one after the other, with the same pattern: the primal scream that woke my mom in the next room, the convulsions, and then a brief, lucid paralysis after my body had stopped heaving. Slowly, slowly, while my mom hovered over me, I was able to open my eyes again, and then move my lips, my fingers, my hands, my legs. I was conscious for the third one, the shorter of the two, and out cold for the second. My mom told me about that one after I came to.

Eventually a neurologist explained what I had experienced. The stabbing pain had been what epileptics call an "aura," a sort of sensory warning shot from the brain before it all goes haywire. An aura could be a burst of light or colors, or a sound, or a sudden smell, like burning toast. Mine happened to be excruciating, but that had its uses: My screams meant that there would always be an adult awake and prepared to call an ambulance if my convulsions failed to stop within a minute or two.

I guess I was lucky in that way. I was lucky, too, that I only ever seized at night, so I never fell down a flight of stairs or collapsed in traffic. And I was luckiest because I grew out of the disease, as child epileptics sometimes do. The misfiring neurons that were causing my seizures were sidelined as I grew, I suppose, and I was given the all-clear by the time I was old enough to learn to drive. Still, even counting my blessings, those seizures were among the scariest experiences of my life. There was something deeply wrong about feeling my conscious self shunted to a far corner of my own brain, forced to watch my body act out without my permission.

Later, during the decade that I played rugby, I saw a couple of athletes lie convulsing on the field after hits to the head. I looked away, as I still do from depictions of seizures on TV medical dramas. I don't like to watch the thrashing and the twisting, knowing that's how I must have looked to my mom, my horrified audience of one.

The epilepsy left me with something else besides an occasional discomfort while watching *Grey's Anatomy*. For years after that first night, the nightmare and the seizure were linked in my mind, one seeming to have caused the other. There was a time when I genuinely believed that another nightmare could cause me to seize, that standard childhood fare like campfire horror stories or scary movies would trigger the convulsions that could kill me.

I avoided scaring myself at all costs. I was convinced that fear itself could hurt me. It was a child's illogical leap, but there was an intuitive truth at its core. Brains and bodies, nightmares and scary stories, these can't be easily divided into separate categories, like the peas and carrots that must never touch each other on a fastidious eater's plate. Our physical brains and our emotional minds—in other words, the brain cells that caused my seizures, and my own feelings of fear—are inextricably linked. Scientists are only just beginning to understand the ways in which our feelings, fear among them, are products of our physical brains.

In the Bible, God regularly commands his followers not to be afraid. According to Rabbi Harold Kushner, the admonishment to "fear not" appears more than eighty times in the text, directed at Abraham, Jacob, Moses, and each of the prophets, among others. Its occurrence is so frequent that Kushner refers to it as an eleventh commandment. But whether we are religious or not, obedience to that command is usually far beyond most people. The fear response system is built into the human body. Fear is, for most of us, simply a part of being alive.

Not everyone experiences fear the same way: Some of us, it seems, have a surplus of it. And managing that surplus, or at least its effects, has been a matter of medical concern for thousands of years. As early as 400 BCE, the Greek doctor Hippocrates was attempting to provide medical treatment to people whose symptoms we would recognize today as phobias: Men who, as described in a perfect phrase that appears to have survived from ancient times, "feared that which need not be feared." Hippocrates and his disciples treated otherwise healthy people who never went to parties or large gatherings, or avoided groups of other people altogether, so convinced were they that they would be mocked and scrutinized or would somehow humiliate themselves. They saw men who were afraid to leave their houses in daylight, and others who were terrified to go near a cliff edge or a bridge. Today we call these conditions social phobia, agoraphobia, and acrophobia.

Unlike many of his peers, Hippocrates did not believe that fear was injected into us by the gods. He believed, instead, that our neuroses had physical causes—specifically, a buildup of black bile in our brains that created overheating and resulted in fits of baseless fear. He treated his patients with improved diet and exercise, to purge the bile from their systems. If that didn't work, he administered a poison that would induce diarrhea and vomiting, presumably purging the bile in the process.

His diagnoses and treatments had their limitations, but at least he was grappling with the problem. After the fall of the Roman Empire, the curtain of the Middle Ages closed over European science and medicine, and Hippocrates's ideas—among those of many others—were swept aside by the Church. Through those centuries, phobics were often assumed to be possessed.

Then came the Enlightenment and a return to a search for more temporal causes. But black bile was out, and individual experiences were in. In 1649, Descartes wrote,

> It is easy to conceive that the strange aversions of some, who cannot endure the smell of roses, the sight of a cat, or the like, come only from hence, that when they were but newly alive they were displeased with some such objects. . . . The smell of roses may have caused some great headache in the child when it was in the cradle; or a cat may have affrighted it and none took notice of it, nor the child so much as remembered it; though the idea of that aversion he then had to roses or a cat remain imprinted in his brain to his life's end.

This idea, that our long-lasting fears stem from unpleasant early-childhood experiences, is still in play nearly four centuries later.

The end of the nineteenth century and the beginning of the twentieth saw key developments in our understanding of fear. The first came from the Russian researcher Ivan Pavlov, who, while studying dog digestion, noticed that his canine subjects began to salivate not only when their food arrived but also in the presence of their usual feeder. To test his suspicion that he had inadvertently trained the dogs to associate the feeder with the food so closely that they responded to one as if it were the other, Pavlov devised a famous experiment: He began pairing the arrival of the food with the unrelated sound of a metronome, and then, after some repetition of the pairing, he presented the dogs with the sound but no food. On cue, they salivated, responding to the conditioned stimulus with a conditioned response. This process, now known as Pavlovian, or classical, conditioning, would become a cornerstone of modern psychology. And it played an essential role in future studies of fear and phobias.

After the First World War, the American psychologist John B. Watson decided to build on Pavlov's work. He wanted to know if a seemingly natural human fear reaction, such as a child crying in response to a loud noise, might grow to encompass fear in other circumstances, too. His test subject was Albert B., a baby whose mother worked as a wet nurse at a Baltimore hospital. Albert was reported to be an emotionally stable infant. "No one had ever seen him in a state of fear and rage," Watson and his graduate student Rosalie Rayner wrote later. He "practically never cried."

First, they determined that the baby was afraid, as is natural, of sudden loud noises. Then, after exposing Little Albert, as he became known, to the presence of a number of small animals and ensuring that he showed no fear of them, Watson and Rayner began the experiment proper. When the baby was just over eleven months old, one of the researchers presented him with a white lab rat. When Albert touched the animal, the other researcher, positioned behind their subject, slammed a hammer into a long steel bar, producing a loud crash. "The infant jumped violently and fell forward," Watson and Rayner wrote, "burying his face in the mattress."

It didn't take many combinations of the noise and the rat for Albert to learn to associate the two. Before too much longer, Watson and Rayner were able to induce whimpering, tears, and an effort to escape from the rat without the use of the noise stimulus at all. Even more than that, they found that Little Albert also now cried and shrank away from a rabbit, dog, and fur coat he'd been exposed to earlier and hadn't feared. Watson and Rayner had successfully created a phobia, or at least a pattern of fear, in a child where there had previously been none.

It's not known what became of poor Little Albert after the experiment and whether he carried his fear of furry creatures with him for life. But we do know something about the fate of Pavlov's dogs. Years after his initial study, a devastating flood in his laboratory nearly drowned many of the animals. For the rest of their lives, the surviving dogs showed signs of being afraid of water.

Around the same time that Pavlov was inducing his dogs to drool on command, Sigmund Freud was launching the field of psychoanalysis. Where Descartes had pondered the possibility that a child with a fear of cats might have had an early negative encounter with a cat, Freud took a different, more opaque, route. In his 1909 study of Little Hans, a child who had seen a carriage horse's violent collapse in the street and had thereafter been afraid of horses, Freud posited that Hans suffered from a variation on what he called the Oedipus complex. The boy actually feared his father, not horses at all, Freud wrote, and that fear stemmed from the child's sexual attraction to his mother. (It's hard not to wonder what Freud would have made of my own hang-ups. Was my hesitation at the top of that escalator about something more than a small child's sense of vulnerability when stepping onto a large and in-motion machine? Maybe I'd rather not hear his answer.)

Today, Freud remains famous for these sorts of theories, and his influence on the fields of psychology and psychiatry has been enormous. But his career nearly took a very different track. Initially, after he graduated from medical school, Freud worked in neurology. He studied the nervous systems of fish and crayfish and became involved in the debate over how, exactly, our brain cells communicate. Freud argued (correctly, as it turned out) for the existence of a physical gap between neurons, and in 1895 he wrote that "the nervous system consists of distinct and similarly constructed neurones . . . which terminate upon one another." The neuroscientist Joseph LeDoux credits Freud with coining the phrase "contact barriers" to describe the connection points between neurons. "Although these notions were amazingly sophisticated for their time," LeDoux writes, "Freud felt that progress in understanding the brain would be too slow for his taste and so abandoned a neural theory of the mind in favor of a purely psychological one. The rest is history."

Freud wasn't wrong: A lot of significant advances in neurology were still decades away. But luckily for those of us who suffer from phobias and other afflictions of fear and anxiety, our understanding of the physical mechanisms of the brain has advanced a long way since Freud gave up on it.

I REMEMBER WHEN I FIRST LEARNED about classical conditioning. It was sixth grade, around the same time that I had my first seizure. I had been given *Gordon*, the Barenaked Ladies' debut album, on a cassette tape for Christmas. The lyrics for each song were dense, layered with references I didn't fully understand, and the second verse of the fourth track on side A, "Brian Wilson," included a mention of Pavlov's dogs.

I can still picture the evening darkness in our living room, the antique wooden armoire that held our stereo, me popping the tape into the tape deck and then unfurling the liner notes from the cassette case and poring over the lyrics in their tiny type. As I usually did when I couldn't understand something, I turned to my mom, and she explained the basics of Pavlov's work.

It didn't make much sense to me then. Why would you want to make a dog drool for no reason? But I filed the information away, and I still can't think about classical conditioning without hearing lead singer Steven Page's voice singing softly in my head. Call it a conditioned response!

Everything I did that evening—listening to language and processing its meaning, learning and storing new information, forming an enduring memory of the whole event—was thanks to the remarkable properties of the human brain. There's a lot going on up there. But until recently, I took it all entirely for granted. I never thought to wonder, *How does the human brain work? What is actually happening when I'm feeling afraid?*

Our brains receive, process, and transmit information using specialized cells called neurons, or nerve cells. Neurons have two sorts of appendages, or branches, dangling from the main cell body: axons, which transmit information; and dendrites, which receive it. The connections at the gaps between the two—the gaps correctly predicted by a young Freud, where one neuron's axon passes information to another neuron's dendrite—are called synapses.

The brain contains more than eighty billion neurons and trillions of synapses. When a neuron "fires"—that is, when it receives a stimulus and sends that information down its axon to another neuron—it can pass its message along at more than one hundred miles an hour; a good thing, considering that while most axons are microscopic, some measure several feet in length, traversing our bodies from brain to limbs. One neurobiologist has estimated that an adult human's axons could measure a cumulative distance of several hundred thousand miles.

Which brings up another point. We tend to think of the brain as distinct, separate from our bodies, of the wrinkled lump in our skulls as the conductor alone on his stand, directing the orchestra. But our bodies and our brains are intricately linked in every possible way; the brain is no more neatly separable from our bodies than our skulls are from the rest of our skeletons, or our hearts are from our arteries and veins.

Together, the brain and the spinal cord make up what we call the central nervous system. Its partner, the peripheral nervous system, is the entire collection of nerves and nerve cells that fall outside of those two core structures: the axons that carry information and instructions out from the central nervous system to our every muscle, and the sensory neurons that carry information from every part of the body back to the center. It's this system that gathers everything we know about the state of our bodies: hot and cold, pressure, pain, and so on. That sensory information gets passed up the spinal cord to the brain for further processing. It's bringing you the image of these words right now.

In the brain proper, our neurons organize themselves into groups and systems, called nuclei, and form structures to complete certain tasks. One key structure for our purposes is the thalamus (or thalami, since, as is the case for most brain structures, there are two, one in each hemisphere of the brain). It's a sort of gatekeeper, regulating the flow of sensory information from the body to the cerebral cortex. The hypothalamus is another regulatory structure. It works alongside parts of the brain stem, the most evolutionarily ancient, bottommost portion of our brain, in governing our autonomic nervous system, the involuntary, invisible system that regulates our internal organs: heart, guts, lungs, bladder, and so on.

The hypothalamus's neighbor, the amygdala, is a critical structure for any discussion of fear—so critical, in fact, that Joseph LeDoux, who has made a career out of the study of fear, named his science-rock band the Amygdaloids. The amygdala's job, in highly simplified terms, is to receive sensory input and to subject that information to a threat assessment. If it finds a threat, it lets the hypothalamus know that it's time to fire up the autonomic nervous system's "fight-or-flight" response. The amygdala can operate without bothering to check in "upstairs" with the cerebral cortex for permission. It can even be triggered by stimuli of which we are not consciously aware.

Above and surrounding all these smaller structures is the big guy: the cerebral cortex, that heavily crevassed dome of soft matter that will dominate in a child's drawing of the brain. It is believed to be the key to our whole array of higher-order mental functions, from perception and consciousness—in the philosophical sense rather than in the raw sense of whether you are awake or comatose—to skilled movements, memory, and intelligence. It's the reason why we have Steph Curry's three-point shot, Picasso's *Guernica*, and the classic *New Yorker* cartoon of an affable piece of rigatoni saying on the phone, "Fusilli, you crazy bastard! How are you?" It's also the reason why we can carry grudges, nurse insecurities, and stew for years about the mistakes we've made.

So let's say you get scared. What does that actually look like, in physical terms?

Picture something like the vision I experienced in that nightmare a couple of hours before my first seizure. You wake up suddenly in the night. You're alone in your home, but you hear a strange noise in the darkness—maybe a footstep, a door opening or closing. That auditory input is first carried from the ear's receptor cells to a major cranial nerve and then directly to the brain itself. There, it's presented with various pathways and options, but for simplicity's sake, let's assume that in this case we have a straightforward journey from gatekeeper to fear trigger. The thalamus sends an alert straight to the amygdala, which in turn alerts the hypothalamus, and then the sympathetic nervous system is fired up. Messages race along the axons throughout your body, from synapse to synapse to synapse, carrying the news of potential danger to your organs, to your skin. Your heart rate accelerates; its loud pounding seems to fill the dark room. Maybe your breath comes faster and shorter, too, and your flesh prickles with sweat beads or goose bumps. Your pupils dilate; your muscles flood with blood, preparing for action. You feel afraid: a sickness in your gut, tightness in your chest. Fear is a full-body emotion.

The physical reactions—the blood flow, the pupil dilation, and so on—can all be traced at the neural level. But what about the *feeling* of fear, which is distinct, as we know, from the physical fear response. Where does it come from?

For a long time, the working theory held that the feeling came first, in response to the fear stimulus, and then the physical response followed from the feeling. This is what's known as the commonsense, or Darwinian, school of thought. But that was more an assumption than a proven mechanism, and these days it has fallen out of favor. Instead, as science has turned its attention to working out that elusive mechanism more concretely, the neuroscientist Antonio Damasio has come up with an answer that, while provocative, ultimately *feels* right to me. The feeling, he argues in a pair of funny and wise books, *Descartes' Error* and *Looking for Spinoza*, is actually derived from that same menu of physical reactions that we would typically view as accessories of, or adjacent to, our emotions.

For the purposes of his argument, Damasio makes an unusual distinction between "emotions"—by which, in this context, he specifically means the physical, measurable reactions of the body in response to an emotional stimulus, the physical fear response—and "feelings," the intangible expressions of emotion in our minds. That may seem odd, or even nonsensical, but it's a key to his case, so keep it in mind.

"We tend to believe that the hidden is the source of the expressed," he writes in *Looking for Spinoza*. But he argues, instead, for a counterintuitive reversal of that order: "Emotions"—again, meaning the physical reactions here—"and related phenomena are the foundation for feelings, the mental events that form the bedrock of our minds."

All organisms have varying abilities to react to stimuli, from a simple startle reflex or withdrawal movement all the way up to more complex multipart responses, like the description of our physical fear processes above, which are Damasio's "emotions." Some of the more basic responses might sometimes look, to our eyes, like expressions of the feeling of fear, and in fact, the machinery that governs them is also implicated in the more complex processes. (My startle reflex, one of our oldest and simplest reactions, has certainly come into play at times when I've also felt afraid. Hello, raptors in the kitchen in *Jurassic Park*!) But the "emotions" are at the top of the heap in terms of complexity, and as such not all organisms are capable of generating them.

Unlike some of the simpler "fear" reactions in simpler organisms (poke a "sensitive plant," watch its leaves curl up), our emotions can be generated by stimuli both real, in the moment, and remembered—or even imagined. That's the gift and the burden of the human mind. But for now, let's stick with an in-the-moment example, like the noise heard in the night. The fact of the noise is captured by the sensory nerves in the ear and is relayed to the brain structures involved in triggering and then executing a response—that's the amygdala and the hypothalamus again. Now your body is reacting in all the ways described above.

So far, so good? The next step, in Damasio's formulation, is the creation of the *feeling* itself.

We know that our bodies are laced with neurons, and that they not only send out information from the brain, they also receive it. So after the outgoing messages have gotten our hearts pumping, our sweat beading, and so on, a series of incoming messages returns to the brain, bearing all of that information about our physical state. Our brains, Damasio explains, maintain incredibly complex maps of the state of the body, from our guts to our fingertips, at all times. And here's the core of his argument: When the incoming messages bearing news of the body's physical fear-state alter these maps, *that's when the feeling itself arises.*

Your brain learns from your body that your heart is pounding, your pupils are dilated, your goose bumps are standing at attention. Your brain does the math and says, Aha! I am afraid!

In his 1884 essay, "What is an emotion?" the philosopher and psychologist William James wrote,

> If we fancy some strong emotion and then try to abstract from our consciousness of it all the feelings of its bodily symptoms, we find we have nothing left behind, no "mind-stuff" out of which the emotion can be constituted, and that a cold and neutral state of intellectual perception is all that remains. . . . What kind of an emotion of fear would be left if the feeling neither of quickened heart-beats nor of shallow breathing, neither of trembling lips nor of weakened limbs, neither of gooseflesh nor of visceral stirrings, were present, it is quite impossible for me to think.

Damasio picks up where James left off. But he doesn't just draw on Victorian-era philosophizing to make his argument. He also works from case studies and his own research; for instance, the case of a Parkinson's patient in Paris. The woman, who was sixty-five years old and had no history of depression or other mental illness, was undergoing an experimental treatment for her Parkinson's symptoms. It involved the use of an electrical current to stimulate motor-control areas of the brain stem via tiny electrodes.

Nineteen other patients had undergone the treatment successfully. But when the current entered the woman's brain, she stopped chatting with the doctors, lowered her eyes, and her face slumped. Seconds later, she began to cry, and then to sob. "I'm fed up with life," she said, through her tears. "I've had enough . . . I don't want to live anymore . . . I feel worthless." The team, alarmed, stopped the current, and within ninety seconds the woman had stopped crying. Her face perked up again, the sadness melting away. What had just happened? she asked.

It turned out, according to Damasio, that instead of stimulating the nuclei that controlled her tremors, the electrode, infinitesimally misplaced, had activated the parts of the brain stem that control a suite of actions by the facial muscles, mouth, larynx, and diaphragm—the actions that allow us to frown, pout, and cry. Her body, stimulated not by a sad movie or bad news, had acted out the motions of sadness, and her mind, in turn, had gone to a dark, dark place. The feeling arose from the physical; her mind followed her body.

This whole thing seemed counterintuitive to me at first, reversing as it does the "commonsense" view. But then I sat back and really thought about my experience of fear. How do I recall it in my memory? How do I try to explain it to other people? The fact is that I think of it mostly in physical terms: that sick feeling in my gut, the tightness in my chest, maybe some dizziness or shortness of breath. (Appropriately, the ancient Greeks' *angh*, the root of the Latin *anxietas* and the ancestor of modern terms like "anxiety" and "anguish," was originally used to mean tightness, restriction.) The conscious thoughts about how I am feeling—*I am not OK; I am afraid*—are decidedly secondary.

Think about how you actually experience the feeling of happiness, of contentment, or ease. For me, it manifests in the loosening of the eternally tense muscles in my forehead and jaw, in my neck and shoulders. My eyes open wider, losing the worried squint. I breathe more deeply.

Or think about the sheer physicality of deep grief, how it wrecks your body as well as your mind. When I look back on the worst of my grief after my mom's death, I remember it as headaches, exhaustion, a tight chest, a sense of heaviness, and lethargy. I felt sad, yes—sadder than I've ever been—and it was my body that told me how sad I was.

One morning, a few months after my epilepsy diagnosis, I woke up with the memory of another vivid dream lodged in my mind. The dream was straightforward: I'd had a seizure. The usual pain, screaming, and convulsions; the same waking paralysis after the thrashing stopped. It seemed so real that I began to wonder if it was a dream at all.

And here was the thing: I'd been home alone the night before. So if I *had* been screaming and convulsing, there would have been no one around to hear.

My mom and I mentioned the phantom seizure to my neurologist at our next appointment, and she was concerned enough to up my drug dosage on the basis of that dream-memory alone. I never seized again, either awake or in dreams.

Dreams and nightmares are one of the stranger manifestations of our brain's ability to make pictures in our minds. They're a phenomenon that remains somewhat imperfectly understood, although humans have been throwing a variety of explanations at them for millennia. Dreams have been messages from the gods or the ancestors, warnings of danger, or glimpses of the future. Hippocrates and Aristotle believed that dreams could be used as a diagnostic tool, that they were telltale mental symptoms of a physical illness. Aristotle wrote, "Beginnings of diseases and other distempers which are about to visit the body must be more evident in the sleeping than in the waking state." A couple of millennia later, Freud posited that dreams were all about wish fulfillment: Our minds acted out the desires that were not permitted to us in waking life.

These days, we have a pretty specific understanding of the mechanism of dreaming—the how, even if not always the why. "Neurochemical changes that occur during REM sleep prime our brains to not only generate but also trust extraordinary visions," the science journalist Alice Robb writes in *Why We Dream.* In simple terms, the chemicals and structures involved in emotion and memory are activated, while the parts of our brain that deal with reasoning and self-control quiet down. "The result," writes Robb, "is a perfect chemical canvas for dramatic, psychologically intense visions."

Still, even with a basic grasp of the chemical processes involved, we find it hard to let go of our sense of the significance of dreams. It's something about the way they cling to your mind after waking, the way they take command of your mood even after their details have faded away. It feels powerful. Of course, my nightmare that first night didn't actually cause my seizure—but still, for me, dreams and seizures remained all bound up together. The phantom dream-seizure only cemented the connection.

According to the psychologist Richard Wiseman, statistics can explain away most of the perceived potency of dreams, their symbolism and their seemingly predictive powers. Wiseman estimates that between age fifteen and seventy-five, an average person might have nearly ninety thousand dreams over thousands of nights of sleep. "You have lots of dreams and encounter lots of events," Wiseman writes in his book *Paranormality*. "Most of the time the dreams are unrelated to the events, and so you forget about them. However, once in a while one of the dreams will correspond to one of the events. Once this happens, it is suddenly easy to remember the dream. . . . In reality, it is just the laws of probability at work." Would I even remember that one nightmare if it hadn't been followed by my first seizure? Or would it have just slipped away like all the rest?

The thing is, laws of probability aside, dreams actually can affect our waking lives, and they can even have a measurable impact on our health. When our real-life partners or family members play the villain in a bad dream, our negative dream-perceptions of them can persist once we are awake. (A 2013 study found that dreamland "slights and betrayals" from participants' partners lingered into the daytime.) Nightmares have been shown to play a role in triggering migraines and asthma attacks, and even, in very rare cases, heart attacks and other potentially fatal events. Here's Alice Robb again:

> A man in his late thirties—a nonsmoker with no family history of heart disease—dreamed that he died in a car crash and woke up vomiting; two hours later, he was at the hospital describing

> the unbearable pressure in his chest. A twenty-three-year-old woke at six AM from a nightmare in which he was murdered alongside his father and had a heart attack at seven. The early-morning and the final hours of sleep—when REM cycles are longest and nightmares are most intense—is the most dangerous period for cardiovascular patients; heart attacks are most frequent, and most severe, between the hours of six AM and noon.

I'd like, for the sake of my own ability to sleep comfortably through the night, to believe that these kinds of cases are utterly freak events, rare as an asteroid strike or a jackpot-winning lottery ticket. But unfortunately they're not quite so unimaginable. And while the two men in Robb's anecdotes ultimately survived, others weren't so lucky.

In 1980, a medical examiner in Portland, Oregon, contacted the Centers for Disease Control. The examiner had noticed that two recent unexplained deaths seemed to have similar properties. Soon enough, more similarly mysterious American deaths were added to the list, and by the end of the decade, more than one hundred of them had been noted. Here's what the deaths had in common. The dead were mostly male and mostly of Southeast Asian origin—a majority of them, in fact, Hmong refugees from Laos. They had all died in their sleep, had generally been healthy, and autopsies turned up no physiological causes of the deaths. Researchers scrambled to understand, seeking out genetic or cardiovascular explanations. "We drew a complete blank," one medical examiner said at the time. "In each case we asked ourselves what they had died from and the answer was 'Nothing.'" With little to go on, the authorities slapped a name on the phenomenon: sudden unexplained nocturnal death syndrome, or SUNDS.

In 1991, Shelley Adler, a doctoral candidate at UCLA, published a theory about the SUNDS deaths in the *Journal of American Folklore*. Adler was a folklorist, perhaps an unlikely source of insight into a bizarre medical crisis. But, as she explained in her paper, "Sudden Unexplained Nocturnal Death Syndrome among Hmong Immigrants: Examining the Role of the 'Nightmare,'" her training was crucial to the formation of her hypothesis. She'd been taught, after all, to listen to the stories of ordinary people—the kinds of people medical experts can sometimes overlook—and to build a more universal truth from their specific stories.

Her bona fides established, Adler got down to her theory. Across cultures, there is an enduring story, or legend, of an event that she calls "the nightmare": a force, often perceived as an evil spirit, that presses down on its sleeping victims' chests, squeezing the life out of them as they lie helpless, conscious but dreaming. In Hmong culture, the nightmare spirit is called the *dab tsog*. Ordinarily, attacks by the dab tsog aren't necessarily fatal, but Adler argued that a combination of factors had changed that. The Hmong had seen staggering casualties in the war that swept Vietnam, Laos, and Cambodia from the late 1950s to the 1970s; the Hmong, who worked in cooperation with the US's Central Intelligence Agency, died at ten times the rate at which American soldiers were killed in Vietnam, and by the time it was all over, it was estimated that one-third of the Laotian Hmong population had been lost. Then came more danger: death or reeducation at the hands of the victorious Communists or a risky flight across the Mekong River to refugee camps in Thailand. By the time the Hmong refugees made it to the United States, they had suffered trauma, hunger, loss of loved ones on an almost unimaginable scale, and the disorientation and alienation that comes when you are wrenched from tight-knit communities and tradition.

Adler argued that it was the stress of all these social factors, combined with the power of the Hmong men's own belief in the dab tsog, that turned a terrifying nightmare into a fatal event. The images in the men's minds had acted upon their bodies, as we increasingly understand they can do. Their feelings were not neatly separable from their physical existence; their fears were not neatly separable from reality. In some ways, it was almost as though their fears had called the nightmare into reality. It was my fear-caused tumble down the escalator, but on a deadly scale.

Sometimes, on my worst days after my mom died, I wondered if all my worrying about her death, my dread of losing her the way she had lost Janet, had called her fatal stroke into being. I felt as though, maybe, I had summoned the nightmare through my own terror.

Of course, that isn't rational, just my grieving mind making connections after the fact. My childhood belief that a nightmare could trigger my seizures may not have been rational or scientific either. But nightmares, it turns out, *can* come true.

3

a fear fulfilled

We turned off my mother's life support on a Friday. I spent the weekend with family and friends in Ottawa, and on Monday I flew home to Whitehorse. I felt strange, fragile, and exhausted all the time. In between long, deep naps, I walked around town feeling like an invisible alien that might, if anyone noticed my presence, fly to pieces at a human touch. But sometimes, unexpectedly, rage, and the energy that came with it, would bubble up through the lethargy, and I would feel a sudden urge to punch another pedestrian as they passed me on the street or to shout in the face of the airport security agent as they scrutinized my boarding pass. It was a terrifying state to be in. I remember praying to whoever might be listening, *Whatever this phase is, please let me get through it without landing in jail.*

With my worst fear already come to pass, I realized, I had something new to be afraid of. Since early childhood, I had understood the loss of a mother as a life-wrecking event, and now it seemed like that same destruction would be coming for me. Now, I was afraid that I would tear everything down—my career, my friendships, my life. When I wasn't completely lethargic, I sometimes felt crazy, wild, like anything was possible. The professional climber and filmmaker Jimmy Chin once said, in his film *Meru*, that he had always promised his mother he wouldn't die before she did. Once she was gone, he told the camera, there were no longer any limitations on the risks he could take. I felt that same mad freedom now.

I imagined myself taking up ultra-running, channeling all my anger and sadness into grim accomplishment. I imagined myself moving to a beach hut in Thailand, filling the emptiness in my life with full-moon parties and cheap beer and sex with twenty-something European backpackers. "Maybe I'll go to Afghanistan," I suggested to a friend over the phone one night, "and become a war correspondent." "That's . . . an option," she said, sounding worried.

I was worried, too. I made myself a promise not to make any major life changes for at least a year. Everything felt so precarious—everything, myself included, seemed so brittle, ready to shatter.

After a week at home, I flew back to Ottawa again, for a previously scheduled visit: a prelude to a work trip to Greenland and the Canadian Arctic. I had debated canceling, but the trip had been booked for months and, if postponed, would have to be bumped for a full year. As for the extra days in Ottawa, I figured they'd do me more good than continuing to fester in my empty Whitehorse apartment. My dad and stepmom were solicitous; my high school friends brought me quiche; a group of editors at a magazine I wrote for regularly had an edible arrangement delivered to our door. I was still exhausted, and some nights I cried myself to sleep, but mostly I did OK. I functioned.

The exceptions, when they came, were dramatic. One night, three of my old friends took me out for a movie: *Magic Mike XXL*. A comedy about male strippers on a road trip would ease my mind. Right? But before the movie started, the trailers ran, and I found myself staring up at Meryl Streep on the big screen. In *Ricki and the Flash*, she played an aging rock star attempting to repair her fraught relationship with an estranged adult daughter. Watching the trailer was like bumbling onto a land mine. By the time Meryl declared that "sometimes a girl just needs her mother," I was out of my seat and running for the illuminated exit sign and the Cineplex's long hallway, my fight-or-flight response on full blast. I made it to a washroom, locked myself inside its single stall, and was overcome by the kind of sobbing where you feel like you might suffocate and die. I shook and fought to breathe.

After the storm passed, five or ten minutes later, I unlocked the bathroom door and crept back into the theater. The strippers stripped, and the rest of the night passed smoothly.

I flew to Greenland. I boarded a small cruise ship, notebooks and camera in hand, ready—I hoped—to work. I met my roommate for the next twelve days, a woman who was on board to celebrate her sixty-fifth birthday, and found myself telling her that my mother had died less than three weeks earlier. I guess I wanted to warn her, to let her know I might not always be the most stable bunkmate. She was understanding. She'd lost her mother at the same age, she told me. She promised to leave the room if I ever wanted solitude to scream or cry or throw things. "You are not yourself right now," she said.

A couple of nights later, I was spilling my sad story at the ship's bar. The bartender was a young woman around my age, and she reached over and held my hand hard. When her mother had died, a couple of years earlier, she told me, she'd gained weight. Her hair had fallen out in chunks. She'd had dreams, so many dreams. I needed to go easy on myself, she said.

That night, I dreamed about my mom for the first time since she'd died. In the dream, she called me on the phone. I answered, and we chatted briefly—I don't remember about what, nothing momentous. And then, as we said our goodbyes and hung up, my dream-self remembered that she was dead, and, in my dream-self's confusion and fear, I was wrenched awake.

A few nights later, while the ship rocked gently in a sheltered strait somewhere high in the Canadian Arctic Archipelago, I dreamed about her again. This one was a nightmare, weirdly reminiscent of the one that had preceded my first seizure, twenty years earlier. This time, I was alone in a roomy, suburban house. In the dream, I knew that it was my mother's house, although in the waking world, my mom never lived in any such place. I knew, too, that my mom was dead, and that looters were coming to her home to steal all her belongings. In the strange logic of dreams, I accepted that it was my responsibility to protect them.

Again, I moved through dim, shadowy hallways, through a landscape of greys, sensing a hostile presence in our home. But in this dream, I carried a baseball bat, and while I felt afraid—as I had in the childhood nightmare—I didn't feel so vulnerable now. Instead, I remember a harsh determination, a violent energy surging in my arms and through my body. *I was going to show those fuckers.* I remember crouching behind a carved banister at the top of the stairs, flexing my hands on the shaft of the wooden bat, prepared to defend the second floor of the strange, empty house with my life. I woke up to a mixture of anger and dread.

Back in Whitehorse after the cruise, I collapsed. For weeks, I hardly left my apartment. I ate delivery Chinese food and frozen meals. I didn't cook, or exercise, or work. I kept the curtains closed and lay on the futon in my darkened living room, watching season after season of bad television. ("I've been self-medicating with police procedurals," I told my dad during one of our now more frequent phone calls. He answered, "There are worse things you could be self-medicating with.")

I had begun to worry about him, too. I hadn't really before; he had always seemed so sturdy, so permanent. He didn't invite my concern the way my mom had. Now I wondered if I would simply substitute my fear of my mother's death for a fear of losing him instead. Maybe, I thought in my worst moments, life from now on would just be a series of fears and losses and fears, one following the other. Maybe adulthood was just an accumulation of sadnesses.

My grief counselor asked me if I thought I was becoming fixated on the idea of my dad's death. I didn't think so, I said. But I felt raw, and I was afraid that another loss, coming too soon, would push me over some unseen but palpable edge.

In mid-October, three months after my mom's stroke, I packed up my car and drove south. Movement, I thought, was what I needed to get me off my futon and out of the inertial haze of grief. It took me three days to get to Montana, where I saw some friends, and camped and hiked in Glacier National Park. I drove on to Missoula, and then Livingston, where I stayed in a funny old motel that shook whenever a train rolled by. Then I carried on south, to Yellowstone, and hiked and camped alone there, too.

I'd always found spending time in the wilderness to be healing, and I've never minded being on my own in the outdoors. But this trip was different. I was too sad, too wrecked, and too tired. I made an effort, but nothing felt good, nothing fed me the way it was supposed to. In Yellowstone, I sat on the side of a bridge and stared down into a rocky gully. I tried to enjoy the view and the quiet, tearing chunks of pita bread into tinier and tinier pieces in my lap. I wasn't hungry. I wasn't soothed by the park. I just kept looking at the drop and thinking, *I don't know how much longer I can be this sad.*

I changed my tactics. I drove out of Yellowstone and hundreds of miles east, to a friend's place in Laramie, Wyoming, where I ate pronghorn tacos and met her friends in a loud cowboy bar. From there, I jumped south to Colorado and another old friend's place. Alone time, I realized, was not going to help me right now. Unsettled by my own mind and where it had gone as I sat on the bridge, I was desperate for company. I lined up coffees, lunches, and drinks wherever I could. After three months of retreat, other people were my lifeline.

A writing residency in Banff, Alberta, the original excuse for the road trip, was a reprieve. For three weeks, I ate food that was prepared for me, and every day some hidden staff member made my bed. I made friends with my fellow residents and sought out their company for meals and hikes. For the first time, I began to feel like a person again—like a human, made of bone and tissue, instead of the porcelain alien I had felt myself to be for so long, hiding among humanity but secretly different and strange, and always on the brink of being smashed to pieces.

The residency ended, but I didn't feel ready to return to my futon quite yet. I drove farther south, to Spokane, Klamath Falls, and then Los Angeles, where I met up with two friends and drove out to Joshua Tree National Park. Camping and hiking were safe enough with company. After we got back to the city, I stayed by the beach, soaking up sun, feeling myself heal.

As the time passed, I considered what I knew about my mom's life. The facts had always made me sad—I'd felt sorry for her, I suppose, from the safety of my own happy existence. But now I saw things differently. I was filled suddenly with overwhelming admiration for her. She had endured so much, and come out of it still so kind and strong, full of love despite all her losses. It was incredible, really, that she had been able to love me so unconditionally, without ever lashing out from her own pain. She had always been afraid that growing up motherless would limit her as a mother to me, but now I realized that her fears were baseless. I wished I had told her how amazing she was. I wished I had appreciated her strength more, instead of dwelling on her sadness.

Recovering wasn't as easy as a walk on the boardwalk, of course. One night, my friend Jim, my host in Santa Monica, suggested we watch the movie *My Neighbor Totoro.* I'd never seen it, and his daughter was a fan. But the story of the sick and ailing mother, her children missing her and worrying for her, was too much for me. It was Meryl Streep all over again. I tried to control myself, but the more I tried to stifle my rising sadness, the more I panicked. My heart rate accelerated, my chest tightened, and the little girls' crying on-screen seemed to reach out and choke me. I gasped out that I needed to stop the movie, trying desperately as I did so not to scare Jim's child with my outsized reaction. The power of my own grief frightened me, and I worried I might somehow infect her with the terror of loss.

Eventually I turned toward home: friends in Santa Barbara, friends in Seattle, and then the long, lonely drive north again. The days were short and cold, and I took it in easy stages, working my way back to the Yukon just a few hours at a time. I made it home in time for Christmas. I had decided in the summer that I would allow myself to wallow until New Year's. After that, it would be time to pick myself up again.

Over the months, I'd found a sort of community. As much as I could, I had sought out friends who had also lost a parent well before they reached middle age. ("Welcome to the dead parents club," one of them deadpanned.) They understood my situation in a way that others didn't, I felt. They didn't say things like "We all have to bury our parents eventually," or "At least it was quick." And they offered me a way forward: Some of them had told me that, as hard as it might be to imagine from the bottom of the well of my grief, I would come out of my sadness eventually, and be a better, stronger, wiser person for it.

I tried to hold on to that promise. I was afraid I would be sad forever; it was hard to imagine being anything but sad ever again. I imagined getting married one day without my mom's presence in the front row: bittersweet, at best, no matter how happy I might be about my choice of spouse. Childbirth without my mom to talk me through the pain and fear and uncertainty? Unimaginable.

One night, talking on the phone to a friend across the country, I said, "I feel like everything in my life will always be at least a little bit sadder now." She had lost her dad a few years earlier, and she didn't try to talk me out of the idea. "Yes," she said. "It will."

I realized that the future I was contemplating was the one my mother had lived through. She'd been married without a mother or a father in attendance; she'd faced the prospect of childbirth without her mother's steadying presence. *She must have been so scared*, I thought. But she'd done it. Her deep sadness had always been inscrutable to me, but now I understood it better. In a strange way, her death had brought us even closer than we'd ever been. A part of her experience that I had never grasped, a crucial element of her life, was accessible to me now. In losing her and grieving for her, I was getting to know her.

Still, as time passed, I realized that my understanding of her sadness would always be imperfect. After all, I had advantages that she never had. By the time I had made it through my first Christmas without her, I knew that in the end I was going to be OK, even if my life was just a little bit sadder. It was like I'd said in a eulogy at her celebration of life: To the very best of her ability, she had never let her sadness touch me. I'd been loved and supported unconditionally by her for more than thirty years, and so I would never really understand the feelings of abandonment and confusion and hurt that laced her own grief over the loss of Janet. I was more resilient than she had been—not just because I was older, or because I wouldn't be shipped off to a series of boarding schools, or because I still had my dad, but also because she had made me that way.

It turned out that my long-held fear, that in losing her I would be broken in the same ways she had been by Janet's death, had been misplaced. Even though I missed her every day, I found some peace in that.

GRADUALLY, MY FEAR of a new loss, another death in my life, began to fade. Where at first the threat and the lesson of my mom's death had seemed to be "You can't handle this; this will ruin you," now I knew my own resilience. I had been afraid I would go down that path my grief counselor had asked me about, that I would develop an unhealthy fixation on the potential deaths of my remaining loved ones. That, having lost my mom, I would live my life in a kind of perpetual cringe, bracing for the next piece of bad news. That fear receded. There would be more sadness in my future, no doubt, but I didn't dread it in the same way I once had. In the end, my mom's death had not only brought my fear to fruition, resolving the nightmare by bringing it into being; it had also taught me to be less afraid.

I started the new year feeling sad but newly empowered. I had faced my worst fear, and I had survived.

Soon after, I began to wonder what other fears I might be able to face down. And following the humiliation of my meltdown on the Usual in February, I moved from pondering to determination.

The three main pillars of fear in my life were my seemingly lifelong fear of heights; my terror of driving, more recently acquired after a series of car accidents; and the fear we all carry, to one degree or another, of losing the people we love. That last fear, I figured, I had made a kind of peace with for the time being. But could I cure, or conquer, or overcome, or at least renegotiate my relationship with the other two? It was time to find out.

PART TWO

4

free fall

In the last moments before I climbed into the Cessna, I turned and faced a young, bearded man who was pointing a video camera at my face. I wore a jumpsuit made of panels of fluorescent orange and green fabric, the colors faded by years of sun and wind. A pair of goggles and a leather helmet were strapped on my head.

"Why are you here?" the man asked.

I took a deep breath. "My name's Eva," I said, speaking to the camera lens, "and I'm here to face my fear of falling from heights."

The small crowd that had gathered around me oohed and cheered as I crawled into the tiny plane, awkward in my elaborate harness. Only the pilot had a seat—all the others had been removed—and I sat on the floor behind him, facing backward, spooning with my tandem dive-master, Barry. Another pair climbed in beside us: dive-master Neil and his charge, Matthew, a first-time skydiver like me. They sat by the open doorway, and Matthew and I bumped fists as the little Cessna rattled its way down the gravel runway. Matthew looked elated. I knew I was supposed to be excited, too, but I couldn't get there. For the moment, I existed in a bubble of cold calm. That, I figured, was preferable to the likely alternative: wild, hair-tearing panic.

Four months had passed since my attack of fear and panic while descending from the Usual. Since then, I had started working full-time again, had started exercising and seeing friends regularly. I had even moved to a new apartment, turning my back on that dimly lit living room, its curtains drawn and futon beckoning. Now the Yukon summer and its endless sunlight was upon us, and I finally felt ready to start on the mission I had set for myself back on that lonely day in February. I had vowed, as I sat alone in the hotel by the side of the highway, that I would learn to understand, and then control, my fear. It was time to get started.

I was acting on a very popular idea, one that's embedded in our culture: The notion that facing one's fears is the key to conquering them. In their third year at Hogwarts, Harry Potter and his classmates are taught by Professor Remus Lupin to face down their fears by laughing at them. In *The Sound of Music*, the abbess tells Maria she must confront her feelings, not hide out in the abbey. And in the novel *Dune*, in the iconic Bene Gesserit "Litany against Fear," Frank Herbert wrote, "I will face my fear. I will permit it to pass over me and through me. . . . Where the fear has gone there will be nothing. Only I will remain." Fear, Herbert wrote, was the mind-killer. I wanted my mind to live.

I'd arrived at the small airstrip in the village of Carcross, Yukon, several hours earlier. Carcross is an hour's scenic drive south of my home in Whitehorse. Among its few claims to fame is the Carcross Desert, billed as the world's smallest, a tiny collection of soft, rolling dunes surrounded by snow-etched mountains and boreal forest. Every summer, a skydiving outfit based in British Columbia caravans up here for a couple of weeks and offers Yukoners the chance to jump out of a plane, plummet through free fall, deploy a parachute, and eventually land in the forgiving embrace of our tiny patch of sand.

The pro skydivers live by the airstrip, just outside the village, for the duration. The vibe of their encampment is somewhere between summer weekend campout and itinerant circus troupe. They gather in a jumble of tents, U-Hauls, cars, RVs, and trucks loaded with campers. Barry is their patriarch. When I met him, he'd been jumping for thirty-nine years, including more than two thousand tandem jumps with clients. He had grey hair and a grey mustache, a big belly, and a bigger voice. He's not what you picture when you think "professional thrill-seeker," but I found his age and experience more comforting than any young gun could have been. As they say in Alaska, there are old pilots, and there are bold pilots, but there are no old, bold pilots.

Randy was another older guy, around to help pack parachutes and roll the Cessna on and off the runway. Kelsey was a young woman, mid-twenties maybe, who'd gotten hooked on jumping and was earning her keep—and her jumps—by handling all of the little company's administrative duties. A tent beside the airstrip served as her office. Bespectacled, bearded Jeremy was the AV guy; he made sure that clients got their photos taken on the way up and the way down, and that their free fall was caught on a GoPro camera. Two more young guys, both named Cody, were trading parachute-packing duties for jump time. Skydiving is an expensive sport, and I gathered that most of the people here were swapping their labor for time in the sky.

When I pulled up, just before 10:00 AM, most people were gathered in camp chairs around a fire. I was invited to sit down, offered tea and a hunk of hot fry bread. I met Barry's crew, my fellow jumper Matthew, and an older couple from Whitehorse—the husband would be jumping today, but the wife planned to stay on the ground. Nothing much was happening for the moment, because the pilot had gone into town to get gas. Under an awning, a flat-screen TV played a looping video of past skydive clients shrieking and grinning and falling, while The Prodigy's hit song "Firestarter" blared. I tried to avoid even glancing at the screen.

I was here because my three most potent physical fears were of heights, speed, and falling. And there was nothing, I figured, that combined all three as effectively—or as horrifically—as skydiving. My notion was to take a blitzkrieg approach to facing my fears. I would force myself to do the scariest thing I could think of, in a full sensory assault on my fear response, and if I came out the other side, I would be . . . changed, right? Empowered. That was the idea. So far, I just felt sick and scared.

Barry introduced us first-time jumpers to the gear we'd be using, how the various safety mechanisms worked, and informed me that if I tried to grab on to the plane as we jumped, latching on in a last-minute panic, he would break my fingers to release my grip if he had to. His tone suggested that it wouldn't be his first time doing so.

Then Kelsey had me sign the bluntest waiver form I'd ever seen. "Sport parachuting is not perfectly safe," it read. "We can not and do not offer any guarantees. We do not guarantee that either or both of your parachutes will open properly. We do not guarantee that individuals at SkydiveBC North or Guardian Aerospace Holdings Inc. will function without error. We do not guarantee that any of our backup devices will function properly, and we certainly do not guarantee that you won't get hurt. You may get hurt or killed, even if you do everything correctly."

The form did nothing to calm me down. I signed my name and handed it over. With the paperwork completed, there was nothing much left to do but wait my turn, and stew.

A little after noon, the first planeload went up: four solo jumpers, including Kelsey and the two Codys. I couldn't watch. Panic was starting to simmer inside me, and it got worse when I learned there were still two planeloads to go ahead of me. I couldn't wait here all day. I couldn't take hours of this grim, sickening anticipation.

Kelsey, back from her jump and still in one piece, found me on the verge of tears, on the verge of getting in my car and fleeing. She tried to talk me down. She was kind and warm, but she had no idea how I felt. Nothing made her happier than jumping out of a plane; she had arranged her whole life around doing so as often as possible. She was convinced that if I would just give skydiving a chance, I would love it. She talked rapturously about the rush of free fall, and then the gentle, blissful float down once your chute had opened. She was sweet, but she was crazy.

She did, however, get me shifted onto the manifest for the next planeload.

With the Cessna back on the ground between groups, Barry showed me how we would enter and exit. The plane was tiny, and when we launched ourselves through its low doorway, we would be harnessed together. There was a careful protocol to follow. I'd pictured us stepping out of a full-height doorway, or even a yawning garage-style opening, like in the movies. But the small plane, plus our joined bodies, demanded an awkward crouch-and-roll. For some reason, the sheer impossibility of the maneuver—really, I was going to tandem-somersault out of a tiny opening in mid-flight?—calmed me down. This couldn't be real. It seemed like a joke.

Then, suddenly, it was time. I pulled on my fluorescent jumpsuit, helmet, and goggles, and let Kelsey and one of the Codys—the taller one, Hollywood handsome, with Jim Morrison locks and an *Easy Rider* mustache—cinch me into my harness. I faced the camera, declared my intentions, and climbed into the plane.

Soon we were airborne, rising up above the desert, and Carcross, and Bennett Lake stretching away into the mountains. The landscape below me was familiar, comforting. Countless times, I had hiked it, biked it, paddled it, driven it, flown over it in commercial jets. I've never minded flying; it was the falling I was worried about. I tried to breathe deeply and focus on the scenery. There was the train bridge. There was the beach. There was the highway leading home.

Somewhere on the way up, shivering with cold and fear, I noticed something: I wasn't sweating. I had expected to be clammy with fear-sweat, but instead I was bone dry.

Perspiration was on my mind because I'd recently heard about a scientific study that used the sweat of first-time skydivers to answer a single question: Can humans smell fear?

We've long known that animals can "smell" fear on each other, although in casual, nonscientific conversation, we tend to talk about it in terms of predators smelling fear on their prey. That's a misunderstanding of the phenomenon. What happens is that prey animals unknowingly emit what are known as alarm pheromones, airborne chemical cues intended to silently warn other members of their species, alerting them to nearby predators and other potential dangers.

Several studies have pointed to the possibility that humans, too, can signal their fears to one another by chemical means, through our sweat. Two of those studies showed that test subjects were able to distinguish between the sweat of a person who was watching a scary movie and a person who was watching something non-frightening. Another found that subjects who had smelled the sweat of scary-movie-watchers demonstrated improvement in a word-association test, suggesting heightened cognition in the presence of a potential threat. Still more studies found an increased startle response in people who'd been exposed to someone else's fear-sweat and an increased likelihood to perceive facial expressions as fearful or negative. The takeaway was clear: People who had smelled another human's fear-sweat were primed for a fear response of their own.

But those studies were all based on observed behaviors. A team of researchers, led by Lilianne Mujica-Parodi of Stony Brook University, wanted to look deeper. "We set out to determine whether breathing the sweat of people who were emotionally stressed produced, in a group of unrelated individuals, neurobiological evidence of emotion-perception," Mujica-Parodi and her colleagues wrote in a 2009 article in the journal *PLOS One.* They decided to use an fMRI scanner (which tracks blood flow to measure brain activity in real time) to determine whether exposure to human fear-sweat provoked a measurable reaction in another human's amygdala, that key brain structure that triggers our fear response.

"We kind of set out to do the first rigorous test of whether human alarm pheromones existed," Mujica-Parodi told me. The team started by collecting sweat from 144 people who were participating in a first-time tandem skydive. Then they used those same 144 individuals as their own controls, collecting their sweat after they'd run on a treadmill for the same length of time that the skydive had lasted and at the same time of day. "Because the tandem-master controlled the descent," the researchers wrote later, "the skydiving condition produced a predominantly emotional but not physical stressor for our sweat donors, while the exercise condition produced a predominantly physical but not emotional stressor." (They confirmed the first-timers' emotional stress by testing their levels of cortisol, a hormone released by our adrenal glands in connection with our fight-or-flight response. Sure enough, they had spiked.)

Then came phase two: presenting the sweat samples to test subjects, using fMRI scans to view how their brains reacted in real time. The fear-sweat, Mujica-Parodi and her colleagues wrote, "produced significant brain activation in regions responsible for emotional processing. . . . Behavioral data, our own as well as those from previous studies, suggest the emotional processing may be specific to enhancing vigilance and sharpening threat-discrimination."

They had shown that when a subject inhaled sweat taken from a stressed or fearful person, their amygdala was activated. In a secondary procedure, they had also shown that what was happening wasn't about *smell*, exactly. Our noses can't distinguish between fear-sweat and regular, everyday exercise-sweat, but our brains react differently to the two. That's what's known as a chemosensory reaction: The pheromones in the fear-sweat trigger our emotional, not our olfactory, sensors.

Then they took it one step further. The researchers hooked another group of test subjects up to an electroencephalogram (EEG) machine, which uses electrodes fixed to the subject's scalp to record patterns of electrical activity in the brain. Basically, an EEG lets researchers see which parts of the brain are reacting to a given stimulus. Once they were wired and ready, the subjects were exposed to both fear-sweat and exercise-sweat while being shown a range of images of human faces, with a carefully manipulated spectrum of expressions, ranging from neutral to angry.

The results were striking. When they were inhaling the exercise-sweat, the subjects' brains only reacted strongly to the angry faces, treating them, but not the neutral faces, as potential threats. But when they inhaled the fear-sweat, subjects reacted strongly to the whole range of faces, from those with neutral expressions to ambiguously angry to clearly angry. The suggestion, the researchers wrote, was that the fear-sweat triggered the brain to create a sort of heightened vigilance in the subjects, a greater attention to the environment around them.

That finding firmed up, and further codified, the results of the previous studies showing that fear-sweat could increase what's known as "defensive startle"—our jumpiness, basically—in test subjects. We can, indeed, "smell" fear on each other. And that chemical alert system prepares our brains to react to incoming threats.

When we spoke, I asked Mujica-Parodi why she had chosen skydiving as a way to gather the fear-sweat she needed. "Skydiving was a way to induce actual danger in a way that was also ethically sound and also scientifically sound," she told me. "Most dangerous situations, like an earthquake, or combat, they're very uncontrolled. So first of all, people have different experiences even within that same context. Even within combat, there are people who have near misses with an IED and there are other people who never get near it. And so trying to take people in those naturalistic environments and see how they respond turns out, scientifically, not to be a very good way to proceed.

"The nice thing about skydiving is . . . it's an experience unlike anything you've ever encountered before. Evolutionarily there's no animal that enjoys the feeling of being dropped, and it's also highly controlled."

The team was able to standardize factors like the height from which subjects jumped and the amount of time they spent in free fall and in descent after their chute was open. And they were able to fit the participants with biometric sensors, so that their physiological responses to the jump—heart rate, and so on—were recorded in real time.

I asked Mujica-Parodi if she'd ever gone skydiving herself. I half expected, given the role the sport had played in her work, that she might be an enthusiast, even an evangelist, like Kelsey.

Yes, she told me, she had tried it. She had a personal principle to uphold: "I don't do anything to my subjects that I don't do first to myself." But she was by no means an enthusiast.

"I'm just not at all a fan of heights," she said. "Actually, that's putting it mildly. I'm very, very phobic. But in the interests of this ethical principle, I did force myself to jump, and I felt very nauseous. . . . And then afterward, I had some dreams, you know, some nightmares. I wouldn't say that it was a trauma in my life—I don't think about it anymore—but I would not say that I enjoyed it."

THE ASCENT TO ten thousand feet seemed to take hours, and as we climbed, the weird out-of-body calm I'd felt on takeoff seeped away. It was like coming out of shock, losing that numbed protection and feeling the full pain of an injury for the first time—only instead of pain, I felt a terror that rose through my body until it reached my lungs and my throat and my brain and threatened to choke me.

Barry, behind me, sensed my growing tension—no surprise, since we were pressed together like a pair of lugers on a sled. He squeezed my shoulder periodically and pointed out landmarks below. As we neared jump height, the Cessna circled around a large cloud, skirting its edge.

"You might be a lucky girl and get a cloud jump," Barry said.

I did not want a cloud jump.

The pilot announced that we were nearly in position for Neil and Matthew's jump. They shimmied toward the gaping hole where the plane's door should have been and nudged themselves awkwardly into a spooning crouch on the lip of the doorway.

Seeing them inch toward open space was nauseating, and I looked away. I couldn't watch them vanish into the sky; I stared at the plane's riveted metal wall instead. The pilot dipped the plane slightly to the right, tipping Neil and Matthew out the door, and then, liberated of their combined 270 pounds, the Cessna sprang back suddenly to the left. My stomach clenched and jerked and I swallowed hard.

Now it was our turn. Barry directed me to roll over and scuttle into position as the pilot got us lined up for another jump. My breath came fast; I struggled for control. I desperately wanted to shout, *No, no, I changed my mind. I don't want to do this.* I clenched my jaw. I knew that if I said the word, they would take me back down to the ground, keep my money, and let me walk away. The whole day would be for nothing.

Eventually I got myself in place, hunched over with my kneecaps level in the front of the doorframe, Barry behind me. I tried to unfocus my eyes so I couldn't see the opening and the endless air next to me, the ground far below. Over the roar of the wind and the plane, Barry shouted last-minute adjustments to the pilot, getting us lined up just right. "Give me five left! . . . Five right!" The seconds stretched out while I fought the urge to quit. I had the sensation of trying to hold up some massive weight, my strength ebbing away, moment by moment.

Finally Barry put his right foot out on the narrow metal step fixed to the plane's fuselage below the open doorframe and yelled for me to do the same. It took me three tries—the wind first blew my foot behind, then in front, before finally I lodged it against his. Next I had to scooch around so my left knee pointed out over the lip of the doorway and lock both my hands onto my harness, gripping a pair of handles at shoulder height. I was glad to have something to hold on to. Ever since Barry had promised to snap my finger bones if need be, I'd had a recurring vision of myself reaching out in panic as we exited the plane and fastening on to the doorframe or a strut with a vise-grip fueled by fear, pulling the Cessna off balance and risking everyone's lives.

We were halfway out of the plane, perched on the very edge of jumping. I was past the bail-out point now. I closed my eyes and tried not to hyperventilate, tried not to think about what was coming. All I could do was stay limp and trust Barry to get us in the air—actually participating in our exit from the plane was beyond me. I felt him rocking back and forth to get our momentum up, heard him yell something, but I was deep in my own head. Then we rolled out of the plane and into space.

Kelsey and Barry had both urged me to keep an eye on the Cessna as I somersaulted out of it. Watching the plane appear to fall away from you when you were the one plummeting was, they assured me, one of the coolest parts of a jump. But I had no desire to watch the earth and the sky spin around me. I kept my eyes shut hard until I could feel that Barry had stabilized us in free fall.

I felt him tap me on the shoulder, then again, and yell something in my ear, and I peeled my hands off the harness handles and thrust my arms out wide like I was supposed to. I tried to think about arcing my body into a slight bow: feet together, head up, my belly pointing the way down. I stared at the ground rushing up at us, and suddenly I opened my mouth and spoke for the first time since we'd started the flight up.

"Holy shit!" I yelled, and the wind seemed to tear the words out of my mouth to make room for more. "Holy shit! Holy shit! Holy shit!!" A small part of my brain noted, amazed, that I could even hear myself, could even produce audible speech, with the force of the air roaring by me. (Later, I would learn that we had reached a peak speed of 101 miles per hour.)

I screamed those same two words over and over through our entire thirty-seven seconds of free fall. Once I got started, I couldn't seem to stop. My voice got hoarse, my throat raw. I kept hollering. Dimly, over the sound of my own swearing, I heard Barry say something about our chute, and then a force seemed to pluck at us from above—not a hard jerk, but now my feet were dangling below me and I could feel my weight pushing down on the crotch straps of my harness.

I stopped yelling. Barry reached forward and offered me the straps that controlled the parachute, to let me steer. It took me a couple of tries to put my shaking hands through the loops, and I was too weak to pull effectively. I could feel him pulling the cords for me from above.

Kelsey and the others had described the long, leisurely parachute descent after free fall as "relaxing." But I couldn't relax—I was too aware of my weight in the harness, my feet dangling, the familiar landmarks far below me. There was the train bridge. There was the beach. There was the highway leading home. Barry spun us around, and I felt sick, hated him for a moment, and quavered that I didn't like that. The fall went on and on. Finally we neared the desert, and Barry took over steering entirely, reminding me of my role in landing.

He twisted us from side to side, tacking like a sailboat to shed speed as we came in over the dunes. Then he gave me the signal to pull my knees up (I did my shaky best) and pull down hard on the chute straps. I braced for impact, but my feet never touched—suddenly I was on my belly in the sand, Barry on top of me. He unclipped the right waist clip so he could roll off me as the ground crew approached, cheering, and freed me completely.

The others clustered around; someone helped me to my feet. Kelsey was smiling, proud of me. I tried to smile back, but my cheeks and lips felt as wobbly as my arms and legs.

"Did you love it?" she asked. She was so hopeful. She had found something that made her world turn, and she wanted so badly to share it, for me to love it the way she did. I wished I had the energy to fake even a sheen of enthusiasm for her, to pretend that she and the others had opened my eyes, had helped me to move through the world more boldly, with less fear.

I was too wrung out to lie. I shrugged and mumbled something vague. *I hated every second of that*, I wanted to say. *I never want to do anything like it again.* But she'd been so kind to me, so instead I stared at the sand and dug around inside myself, trying to find some pride in my accomplishment, some kind of silver lining with which to cover up the apparently bottomless chasm of fear I carried inside me.

Later, after I'd stripped off my harness and helmet and jump-suit, after I'd watched Kelsey and the others jump again, after I'd calmed down enough to safely attempt the drive home, I did find some pride. I *had* done it, after all. I hadn't backed down, pulled the plug at the last minute and forfeited my money and my dignity. I hadn't clutched on to the airplane as we rolled out of it, killing us all. I hadn't screamed the entire way down.

These were small victories. But I knew now that if I was going to achieve a real transformation, to rearrange my relationship with my fears, it would not be through shock and awe. One four-hundred-dollar skydive was not going to solve my problems. I needed to be smarter, more systematic, more scientific.

There was more than one way to face my fears. If necessary, I would try them all.

5

on the wall

The panic grew with every move I made: every time I gripped a small handhold with suddenly sweaty palms, placed my soft rubber-soled climbing shoes onto a small ledge or nub in the granite face. My chest seized up. The fear gripping my lungs and my brain made me dizzy. I breathed loud and fast through my mouth, and my brain screamed warnings at my body. *Stop! Go back! Don't do this! You will fall. You will get hurt! You! Are! Not! Safe!*

I was only a few feet off the ground. It was an early summer evening at the Rock Gardens, a popular climbing crag in Whitehorse, soon after my miserable skydive. And as I'd expected, I was utterly failing to remain calm as I tried to climb.

It was my first attempt at a do-it-yourself regimen of exposure therapy, my new plan to cure my fear of heights. And it was an inauspicious start: In the end, I managed to force my way six or seven feet up a twenty-six-foot route before I begged my climbing partner, belaying me from below, to lower me down. As my feet touched the ground, I tried to control my panting and avoided looking anyone in the eye. I was facing my fear, but it was hard to imagine my resulting feelings, or my control over them, ever improving.

I HAVE NEVER BEEN ABLE TO EXPLAIN WHY I froze at the top of that airport escalator. All I remember is that I didn't feel safe; in fact, I felt certain that I would fall. (And then, of course, I did.) But decades later, I realized that my jolt of fear in that moment, sudden and without explanation, was not an isolated incident. There was a pattern.

Throughout my childhood, I never climbed trees—I just didn't want to, I figured—and I was uncomfortable when my friends and I clambered up to sit on top of the monkey bars in the playground at recess. But if I ever thought about it, I put my nerves and reluctance down to a general timidity: normal scaredy-cat stuff.

Then came my panic as I clung to the mast on that tall ship as a teen. But after I was back on dry land, I put the incident away in my memory; it wasn't something I wanted to dwell on. I never tried to put a label on what had happened, never interrogated the event further.

Years passed. I finished high school, then undergrad, then grad school. After I wrapped up my master's degree, I liquidated what was left of my student loan and went backpacking with two friends in Europe. I'd developed a fascination with the art and architecture of old churches, and we hit cathedral after cathedral across the southern half of the continent. We visited a few cathedral towers, and I gritted my teeth going up and down the narrow stone stairways to reach each one. But I didn't truly lose it until Florence.

I'd made it to the top of the Duomo, as the city's cathedral is known, and was breathing deeply, trying to stay calm and enjoy myself as I looked out over the city's terra-cotta rooftops. The famous steep red dome of the cathedral curved away below me, and as I glanced down at it, suddenly all I could think about was how it would feel to tumble over the flimsy metal railing in front of me, to slide down over those red tiles toward the edge. I could picture it, could feel it in my mind and my body, my speed accelerating as I slid, my complete inability to stop what was coming.

I couldn't breathe.

The viewing platform was crowded with tourists. I pushed through them to the wall and slid down with my back against it, put my head between my knees to block out the view, and hyperventilated through my tears. My friends found me there, eventually talked me to my feet, and held my hands while we inched back down the twisting staircase to solid ground and safety. We didn't visit any more cathedral towers after that.

Still, somehow, "I am afraid of heights" didn't become part of the story I told about myself. Today, the thread seems obvious, but when I was living these incidents, each one seemed unrelated. Never mind the fact that now, when I look back, my queasy visions as I clung to the mast on the tall ship, as I stared down over the cathedral dome in Florence, and as I refused to descend the icy creek seem almost identical—the specific circumstances varying in each incident, sure, but the feelings, the irrational visions of my doom, the same. At the time, I either couldn't see their similarities or, on some level, refused to. Aside from my short flirtation with epilepsy, I had always been healthy, and happy more often than not. I didn't want to label myself as phobic, or even fearful. That kind of thing leaned a little too close to the words "mental illness" for my comfort.

So instead, I quietly practiced what I now know is referred to in clinical circles as avoidance. I quit sailing on tall ships. I stopped scaling the stairs of cathedral towers. I was long past the age of climbing trees or playing on monkey bars anyhow. After my money ran out on that Europe trip, I flew home to Ottawa. I worked a few different jobs before launching my new life as a freelance writer. I moved to the Yukon. And then, living in proximity to mountains for the first time in my life, I was forced to face the truth.

It was May, a weekend, about six months after I'd made the cross-country move to Whitehorse. Some friends and I were on our way to the village of Haines, just across the Yukon-Alaska border, for a craft beer festival. We stopped partway there, at a place called Paint Mountain, so my friends could do a bit of rock climbing. I had no intention of participating, but I figured I'd tag along for the hike and then enjoy the sunshine and the view of the landscape around us while everybody else climbed.

It was an easy approach hike, on a marked trail that climbed gradually uphill from a dead-end road in a rural subdivision. But as we walked up a broad, gently sloping rock face, my foot slipped. Just a little—I didn't fall—yet the slip, the sudden uncertainty, was like a trap door opening in my brain. I plummeted into a full-blown fear response: heart hammering, pupils wide, adrenalin flaring. Again came the wild visions of my own doom. I felt as though I might tumble down the entire mountain, never mind that the slope was too gentle and vegetated for me to get far, even if I tried to roll myself down. I was convinced that a catastrophic fall was imminent. So I did what seemed safest, the only thing I could think of that would minimize my exposure. I lay flat on the rock and then curled up, suddenly fetal.

I'd been hiking last in line. In a pattern that would repeat itself years later, on the Usual, I pretended that my reaction was reasonable and normal. I called out from my prone position, trying to sound calm, just letting them know that I wouldn't be going any farther. I would just stay right here until they were done climbing. My friend Lindsay turned around, saw me lying there, and retraced her steps to crouch beside me. Obviously they were not going to leave me here, she said, concerned but calm. She talked me to my feet, and the whole group turned around and trooped back the way we'd come.

After that, I finally started thinking it, started saying it out loud: *I am afraid of heights.*

Specifically, I realized, as I sifted back through my memories of each major incident, I was afraid of falling from heights. Airplanes were fine, elevators were fine, sturdy bridges and balconies were OK, too. My fear was triggered when I sensed exposure, when I felt like my own feet could betray me and send me tumbling.

Once I saw the pattern, I couldn't understand why I'd missed it for all those years.

Acrophobia, or extreme fear of heights, is among the most common phobias in the world. One Dutch study found that it affects as many as one in twenty people. Even more people suffer from a non-phobic fear of heights; they don't meet the bar to be technically diagnosed, but they share symptoms with true acrophobes. All told, as much as 28 percent of the general population may have some height-induced fear.

Plenty of people work around acrophobia by avoiding triggering situations. But in my new Yukon life, that was almost impossible. Avoiding exposure to heights, to places where I felt like I might fall, meant avoiding hiking, climbing, mountain biking—all the things I was trying to learn to do, all the things my friends liked to do, all the best ways to enjoy my wild new home.

Later that same summer, after the Paint Mountain incident, I went backpacking with a group of friends. We planned to hike the Chilkoot Trail, the classic Klondike Gold Rush route from coastal Alaska over the mountains to northernmost British Columbia and the Yukon. Most people spread the whole hike over three to five days, and the crux of it, which generally falls on day two, is the Golden Stairs, a steep climb up a jumble of fallen rock. If you've ever seen one of the classic photos of gold-seekers bound for the Klondike, silhouetted in a long line as they work their way up a mountain slope, hunched under the weight of their packs, that's the one. It's the last stretch before the summit of the pass.

It was only my second real backpacking trip, and I was easily the slowest of the group. By the time I reached the base of the stairs, my friend Florian was already at the top. Because we all knew the stairs were going to be hard for me—I had, after all, finally identified the pattern and admitted my problem—Florian left his pack at the summit and scampered back down the boulder field unencumbered. When he reached me, where I'd started picking my way upward, fighting the sudden, panicked feeling that the wind was going to blow me backward off the mountain, he took my pack from me. Then he paced me as I climbed, or rather crawled. The only way I could convince myself to keep going, with my mind screaming that the wind would hurl me to my death, was to climb Gollum-like, on all fours. Belly to the ground, I ascended.

The next summer, Lindsay took me out rock climbing for the first time. Even back then, I had the vague idea that facing my fear might help me to control it, and I didn't want to accept that I would simply have to bow out of future excursions with my friends.

With Lindsay coaching me, my chest tight and my breath coming fast, I crept up a rock wall at White Mountain, a popular climbing area about an hour outside Whitehorse. I made it to the top, slapped the anchor in triumph, and was lowered back to the ground again. Suddenly, anything seemed possible. I could do this!

I was halfway up a second climb, stalled out and starting to fret, when a new carload full of climbers arrived below me. These were near-strangers, not the trusted friends I'd come with, and as soon as I heard their voices, I felt my anxiety spike. This, I realized, was another element of my acrophobic meltdowns over the years: They were always worse when crowds of people were gathered around me. It had been that way on the tall ship and in Florence. With strangers watching me, my fear of falling was compounded by my fear of embarrassment, of humiliation, and then it snowballed. (Phobia, meet your unwelcome cousin, social anxiety.)

I begged Lindsay to lower me to the ground before I could whine or cry or otherwise shame myself in front of the newcomers. I didn't climb again that day, or that year.

But I did get a new pair of hiking boots that summer, replacing my worn-out, smooth-soled older pair. The new boots, in a stiff, traditional style and made of leather rather than a mix of modern fabrics, bought me a new confidence on narrow ledges and steep terrain. I trusted them—and, I guess, I trusted myself more when I wore them.

Later that same summer, with my writing career in disarray and my credit card bill mounting, I took a job as a laborer in the Yukon's then-booming mineral exploration industry. Some big finds had kicked off what people were calling a second gold rush, as mining companies rushed to stake and sample territory in the hope of locating significant gold deposits. I wound up in a remote, seven-person soil-sampling camp, a forty-five-minute helicopter ride from the nearest dirt airstrip, and another forty-five-minute flight in a small plane from there back to the nearest village and the highway. My job was to follow a designated line on a map, using GPS to navigate, and to collect a sample at regular intervals. I'd dig down a foot or so, bag a fistful of soil for later testing in a lab, fill out some paperwork, mark the spot on my GPS, and move along.

Simple enough, but the terrain was . . . challenging. On flat ground, we would have had our work laid out on a grid. But since we were in the mountains, sampling dirt from the hillsides, we worked along contours instead. Each of us would be assigned an elevation for the day—4,000 feet, say, or 4,200—and then we would do our best to stick to it. Whatever we encountered, we were meant to cross, sticking to our line, whether that meant slogging through dense brush, side-hilling across steep scree slopes like mountain sheep, or clambering over, across, or around bare rock formations.

The work scared me—the places I had to go, alone, and the things I had to do—but I never froze, never melted down. I got through each challenge, and I got stronger and more confident every day. When I came home after a month, my friends joked that all I'd needed to cure my fear was the right incentive: a ten-thousand-dollar credit card bill on the one hand, and on the other a chance to earn two hundred dollars per day, plus room and board, to start paying it off. It was my own personal, but temporary, miracle cure.

From then on, it seemed like maybe I was getting better, a little. Better at controlling myself, at least, even if I remained nearly as afraid as ever. I was never again as fearless as I'd become during the soil-sampling job, but I kept hiking. I was cautious about which trails I tackled but never again felt obliged to lie down on the ground to save myself. I started ice climbing, casually, in the winters with Ryan and Carrie—the sport was both terrifying and satisfying. For several years, I felt like I was making progress on "my heights thing," as I usually called it. And then came my meltdown on the Usual, and suddenly it felt like all my work had come undone.

I regrouped, determined to try again. The skydive had been my first step, an extreme attempt to force my way through fear. But after that effort failed so spectacularly, I decided to try something simpler, something quieter. So in the summer after the incident on the frozen creek, and a year after my mom's death, I went to the bookstore and bought a copy of *The Anxiety and Phobia Workbook*. Back at home, I flipped to the chapter titled "Help for Phobias: Exposure" and read through the advice, exercises, and worksheets it contained. The book suggested that I could design my own program—that, with a little persistence, I could heal myself.

I decided to build myself a DIY cure, or at the very least a coping mechanism. I would learn to rock climb, putting in real effort and commitment for the first time, and I would use that learning process as a form of exposure therapy. I had climbed, badly and fearfully, a handful of times since that first outing with Lindsay. I hadn't stuck with it, because I didn't enjoy it even to the limited extent that I enjoyed ice climbing. It scared me at least as much, but didn't fulfill me in the same ways. Still, I realized, it was accessible, and it was replicable. I could do the same climbs in the Whitehorse area as many times as I wanted. And it had one more clear benefit: It definitely terrified me. The workbook told me that I would have to be willing to take risks, learn to tolerate discomfort, and persist if my plan was going to work.

I was going to master my fear by exposing myself to it, over and over again.

After my mom died, my dad and I started talking a lot more. It wasn't that we hadn't been close before—growing up, I spent every other week at his house, eating dinner together, listening to public radio, and talking about the news of the day. He taught me to listen closely and think critically, to construct an argument of my own and to find the holes in someone else's. I remember him playing devil's advocate across our dining room table, pushing me to think through my beliefs and positions, to articulate them and defend them. (It drove me crazy, but it did me good.) Later, after I'd moved across the country, we had a ritual whenever I came back home to visit: We'd go out to a pool hall, play a few games, watch some hockey, and talk politics.

But once my mom was gone, things changed, deepened. We'd never been prone to talking about feelings much, and now, with both of us painfully aware of the gaping hole that had been blown in my emotional support network, we made more of an effort. So it was only after my mom's death that I learned that my dad, too, had once been afraid of heights.

This was news to me. I'd watched him climb ladders to do chores and maintenance work my whole life. For a couple of years when I was a kid, after the job that had moved us to Ottawa had dried up, he'd earned a living as a handyman, and I'd seen him scrambling across rooftops on two- and three-story buildings around our neighborhood. But it was true, he told me. He'd been deeply afraid of heights. As a kid, he'd had a recurring dream about falling from the top of a mountain. Then, in the summer after high school, he'd gotten a job at a steel mill. The job required him to climb up ladders and walk across catwalks above the firebrick enclosures where the steel was melted, swaddled in an asbestos overcoat for protection from the heat. Sometimes he had to help repair the enclosures' brick roofs, with the molten metal visible below him. Like me, when I worked in that mining camp, he had, out of sheer necessity, and without entirely realizing what he was doing, enrolled himself in a program of informal exposure therapy. And it had worked.

We both wondered if I had inherited my fear from him. Later, I learned that it was possible. We still don't entirely understand the origins of phobias, the mechanisms for their acquisition. There are various theories, and my bet is that no single answer is *the* answer, that there's no one theory to rule them all.

There's the evolutionary explanation, in which phobias are the lingering result of the reasonable fears on which ancient humans acted to stay alive—once necessary responses but now vestigial, hanging around like our wisdom teeth. "The brain's hardwiring determines how ready we are to become afraid of something," Helen Saul writes in *Phobias*, an overview of the history and evolving science behind the phenomenon. "It provides a kind of mold for our fears." Plenty of specific phobias fit neatly within this explanation: fear of heights, sharks, snakes, tightly confined spaces, the dark. These were all things that could have killed a hunter-gatherer. And the evolutionary view might help to explain why some modern objects that probably should scare us—cars, say, or guns—don't tend to inspire phobias. But it doesn't offer much when you consider social phobia, or agoraphobia. It's hard to see what kind of ancient advantage those fears would have brought to someone trying to survive in a communal, outdoor world.

Then there's the possibility that phobias are heritable, encoded in our genes and passed down through families. One broad study of families, phobias, and anxiety disorders, in New York City, found that the immediate relatives of people who had undergone treatment for a specific phobia were themselves three times more likely than usual to have a specific phobia. Family members tended to share similar, but not identical, phobias, the study found. "Where a parent feared dogs, the child hated cats; or if a girl disliked the dark, her brother feared heights," Saul writes. The categories seemed distinct, too: Specific phobias clustered together within families, but they did not appear to be risk factors for anxiety, depression, or social phobias.

The researcher behind this study, Abby Fyer, argued that specific phobias are passed down genetically in what Saul describes as a "discrete bundle," separate from other, broader anxiety disorders. Other researchers disagree, arguing for an umbrella gene for the whole set of conditions. "Both camps agree that genetics *and* environment contribute to the development of phobias," writes Saul. So far, neither camp has had its view definitively borne out by gene mapping.

Related to the gene theory is the idea that phobias are extensions of our personalities, our essential natures. In the 1950s, the American psychologist Jerome Kagan became interested in how, or whether, people's personalities endured over the course of a lifetime. Following up on a long-term personality study, he checked in on a number of adults who'd been assessed decades earlier. He found that, broadly speaking, the adults had aged into people very different from their childhood selves. But one characteristic remained relatively stable. As Helen Saul writes, "Children who were fearful of the strange and new grew into adults with the same reservations."

Kagan went on to set up his own long-term personality study, hoping to better understand what became known as inhibition, the character trait that sees children shrinking away from strangers rather than displaying curiosity—like when I used to hide in my mom's long skirt as a shy toddler. Later he collaborated with another psychologist, Jerrold Rosenbaum, whose work focused on adults with anxiety disorders and agoraphobia.

Rosenbaum and Kagan gathered a group of children with a family history of anxiety disorders and analyzed them alongside Kagan's longer-term research subjects, who were selected from a population without a history of these sorts of issues. They wanted to understand the links, if any, between inhibition and larger-scale anxiety and phobia issues, and their results were clear. "Where parents had panic disorder and agoraphobia," Saul writes, "their children were more likely to be inhibited than children whose parents were healthy." And inhibited children, regardless of their parents' status, were at greater than average risk of anxiety disorders.

It's always difficult to parse nature and nurture, and there are many more children with an "inhibited" temperament than there are people with social phobias. But Kagan and Rosenbaum's research suggests that, at the least, inhibition in a child is a risk factor for larger issues, whether or not they have a parent modelling anxiety and avoidance for them.

The straightforward idea that phobias are derived from previous upsetting experiences is still on the table, too. Think back to Little Hans and his fear of horses—only instead of the Freudian take, in which Hans subconsciously fears his father's violent retribution for the boy's lusting after his mother, something simpler happens. The boy sees the horse crash to the ground; the boy is terrified by the sight, the sounds, the animal's fear and power; now the boy is scared of horses. Seems logical, right? And it does seem to explain some phobias. But a study in New Zealand, comparing data from children who'd experienced significant falls from heights when they were small and their fear of heights as adults, found no overall relationship between falls and later acrophobia. In fact, kids who experienced "significant injury" from a fall between five and nine years old were less likely to develop a fear of heights. The researchers' suspicion was that it was the children's experience with "safe exposure" to falls that might have protected them from phobia. Perhaps that initial exposure can be tracked back to temperament: Children with a lack of fear, due to their more reckless, uninhibited natures, had the falls in the first place. Afterward, instead of the falls producing a phobia, their fearlessness remained intact.

I've wondered over the years if my fear of heights stems from that incident on the escalator at Pearson Airport. The problem with that theory, though, is that I remember my fear so clearly, and it hit me *before* I fell. Where did it come from? Evolution? My dad's genes? My generally passive or inhibited nature as a child?

There may be no way to know for sure. But researchers have also studied acrophobia on its own rather than attempting a blanket explanation for all phobias, and one study offers a clue. If I'm anything like the subjects of this research, I probably have measurably subpar control over my body's movement through space, as well as an over-dependence on visual cues—which are distorted by heights—to manage my movement through the world.

In other words, I am afraid of falling from heights because I am more likely than other people to fall from heights.

For a 2014 paper in the *Journal of Vestibular Research*, a team of German scientists studied the eye and head movements of people who are afraid of heights, plus a control group, as they looked over a balcony. They found that their fearful subjects tended to restrict their gazes, locking their heads in place and fixing their eyes on the horizon rather than looking down or around at their surroundings. That description will be familiar to anyone who's ever felt afraid of heights or tried to counsel someone who is: *Don't look down. Whatever you do, don't look down.*

According to this study, here's how my reaction plays out: I fix my gaze on the horizon as a defense mechanism against my fear, but because that fear is rooted in my overreliance on visual cues, restricting my range of vision only makes things worse. It's a cycle. My brain knows that my body is bad at navigating heights, so it sends out fear signals as a warning. My body shuts down in response, which only increases the likelihood that I will actually harm my klutzy self. And thus a once rational response to a reasonable concern feeds on itself, growing and spreading to the point where I can hardly stand on a sturdy stepladder.

It's only one paper, one theory among many. But when I think about how I behave when my fear of heights is active, it's a theory that feels true to my experience.

Just as there is an array of theories about the causes of phobias, there have been various theories on how to treat them. Freud and Little Hans led the way into the twentieth century.

The Freudian school of thought held that the act of bringing the subconscious associations and urges that were driving a person's phobic behavior out into the light would resolve the problem.

Then came John Watson, Rosalie Rayner, and Little Albert. Now it was clear that fears could be induced—that the memory of past events could be explicitly linked to later fear responses. Where Freud had emphasized subconscious desires, the new school emphasized the phobia sufferer's actions, their conduct, above all. "Behaviorists saw fears as maladaptive conditioned responses," writes Joanna Bourke in *Fear: A Cultural History*. And there was a solution for that.

Watson and Rayner had always intended to try to undo Little Albert's newly conditioned fear of furry critters. But, so the story goes, Albert and his mother left the hospital before the researchers had the chance. (I wonder why.) So it was left to Mary Cover Jones, then a graduate student at Columbia University, to figure out how to reverse the fear conditioning process.

Soon after Watson and Rayner's groundbreaking but ethically dubious study, Cover Jones and her colleagues studied a toddler named Peter. (Inevitably, like Hans and Albert before him, he became known as Little Peter.) Peter was deeply afraid of a white rat, and he also exhibited fear responses to a white rabbit, a fur coat, feathers, and so on—white fluffy objects and creatures, basically. "This case made it possible for the experiment to continue where Dr. Watson had left off," Cover Jones wrote in a 1924 article in *Pedagogical Seminary*. "The first problem was that of 'unconditioning' a fear response to an animal, and the second, that of determining whether unconditioning to one stimulus spreads without further training to other stimuli."

Cover Jones and her colleagues approached the problem in two phases. First was the "unconditioning" phase, sessions in which a white rabbit (which seemed to scare him even more than the rat) was present while Peter was left to play with other children, none of whom showed any fear of the critter. "New situations requiring closer contact with the rabbit had been gradually introduced and the degree to which these situations were avoided, tolerated, or welcomed, at each experimental session, gave the measure of improvement," Cover Jones wrote.

Peter responded fearfully at first if the rabbit was anywhere in the room. Then, gradually, he learned to remain calm if it was locked in a cage twelve feet away, and then four feet away, and so on.

The second, "direct conditioning," phase was more explicitly Pavlovian. Peter was put in a high chair and given his favorite foods to eat, and the rabbit was placed nearby and then brought closer and closer over multiple sessions. Instead of pairing the rabbit's presence with a negative stimulus, like a shock or a loud noise, it was paired with a positive one instead: the snacks.

Eventually, with some relapses along the way, Peter was able to touch and play with the rabbit.

Mary Cover Jones concluded that Peter "showed in the last interview, as on the later portions of the chart, a genuine fondness for the rabbit. What has happened to the fear of the other objects? The fear of the cotton, the fur coat, feathers, was entirely absent at our last interview."

Cover Jones's work was not initially widely noticed, but in the 1950s, the psychologist Joseph Wolpe leaned on her findings to develop a treatment that he called "systematic desensitization." It involved a combination of relaxation techniques and imaginative work. Patients were taught to enter a relaxed state, loosening their muscles and letting go of their tension, and then, gradually, to imagine themselves being exposed to the object of their fears. The idea, Joanna Bourke writes, was that "relaxation was incompatible with fear, and countered the fear response." The hope was that, with time, the mental habit of relaxation would become dominant, a gentler form of unconditioning than what Little Peter had undergone.

Meanwhile, psychologists weren't the only ones working on potential remedies. The field of neurology was still young, but it was growing, and it included several proponents of procedures that later became known as psychosurgery. The most famous of these procedures is the lobotomy.

We all know the term, but perhaps not the grim details. First attempted in the late nineteenth century, and then adopted more widely beginning in the 1930s, the lobotomy involved the targeted (well, loosely targeted) removal of otherwise healthy brain tissue with the end goal of altering the patient's behavior. Lobotomizing someone wasn't about cutting out a brain tumor, or even the kinds of malfunctioning brain cells that caused my epileptic seizures when I was a kid. It was something else entirely, something that now strikes most of us as invasive, violating, and deeply wrong: an effort to permanently pacify people whose mental or emotional state meant they didn't quite fit in.

"The best method," wrote Walter Freeman, a pioneering and prolific lobotomist, "is just to cut until the patient becomes confused." The neurosurgeons who adopted the practice knew that their patients would be left with an array of impairments, physical and mental. Nonetheless the treatment was applied to thousands of people between the mid-1930s and the mid-1950s. (It also saw a brief resurgence of popularity in the 1970s.) People were lobotomized for all sorts of reasons and ailments, including anxiety, depression, and other "neuroses." Phobias were on the list.

In Joanna Bourke's telling, the lobotomists were under enormous pressure to find a cure for a gamut of mental disorders. In the 1930s, first-time admissions to American psychiatric hospitals were increasing by 80 percent each year. And soon the Second World War would create a whole new wave of distressed, fearful, and traumatized people.

Surgery wasn't the only physical, hands-on option. Doctors also treated phobics and others with a "metrazol storm"—injecting a drug that would induce violent seizures—as well as with insulin shock and electroshock therapy (ECT). While the metrazol treatment didn't achieve a cure for phobias, it did cause spinal fractures in 42 percent of patients. And the ECT could indeed, it seemed, dull a patient's fears, but most often that improvement came as part of a new, more general numbness. Here's the recollection of Stanley Law, a patient who underwent ECT in the middle of the century in an effort to cure his phobia:

> I lay fully conscious on the table, full of trepidation, surrounded by male nurses, insulation was pasted to my temples, a rubber pad was stuck between my jaws, and the electrodes were placed in position; in much the same way pigs were prepared in the slaughterhouse. The low voltage electricity was switched on, I felt the early vibrations, and then I knew no more. Upon

> regaining consciousness, I found myself much as I was before. I was on a kind of table. I didn't for a time know where I was or who I was. Gradually, I saw the mass of equipment around me, vagueness was replaced by a slight awareness. I had some sort of idea that I knew the lady by my side, although I didn't for some time realize that she was my wife. My memory was affected. Part of me wanted to panic now, but I couldn't. All I felt was a benumbing, vegetative, timeless, motionless dimness, a lack of sensory perception, and a startling diminution of the life force.

Law's experience, according to Bourke, was fairly typical. He underwent ECT seven times before his phobia was declared sufficiently "dulled." (I don't know about you, but, all in all, I think I might have preferred Hippocrates's method, the inducement of vomiting and diarrhea to purge my body of its black bile.)

From mid-century on, these sorts of treatments fell out of fashion. The turn away from lobotomies and electroshock therapy is sometimes attributed to a broad cultural backlash—with help from grim depictions of the treatments like the one in *One Flew Over the Cuckoo's Nest*—but the change was also driven by the new availability of less dramatic treatments. In the early 1950s, chlorpromazine became the first drug to be marketed as an antipsychotic. It was followed by the creation of many more drugs aimed at treating the whole gauntlet of mental disorders: antidepressants, anti-anxiety medications, antipsychotics.

Early in his career, Bessel van der Kolk worked on the psychiatric ward that produced one of the first major studies legitimizing pharmaceuticals as a superior alternative to traditional talk therapy. In 2014, in *The Body Keeps the Score*, he looked back with some skepticism at the drug revolution that changed psychiatry:

> Now a new paradigm was emerging: Anger, lust, pride, greed, avarice, and sloth—as well as all the other problems we humans have always struggled to manage—were recast as "disorders" that could be fixed by the administration of appropriate chemicals. Many psychiatrists were relieved and delighted to become "real scientists," just like their med school classmates who had laboratories, animal experiments, expensive equipment, and complicated diagnostic tests, and set aside the wooly-headed theories of philosophers like Freud and Jung. A major textbook of psychiatry went so far as to state: "The cause of mental illness is now considered an aberration of the brain, a chemical imbalance."

A cynical view of his colleagues' motives? Maybe. But the shift was real.

These days, the pendulum has swung partway back again. People coping with phobias and other disorders have a menu of treatment options available to them, and often various components—drugs, talk therapy, other methods—are used in combination. Even ECT has made a partial comeback, in an altered, more humane form.

A FEW WEEKS AFTER THAT FIRST OUTING, I was back at the Rock Gardens. The route I was attempting was a beginner's climb, laughably easy for most people with any experience. And it came with a cheat option: Taking a detour of a few feet to the right, into a wide crack between two rock faces, made it even simpler. But to get to the crack and the easiest way up, I had to make one slightly tricky move. I would have to step forward with my left foot, balance the toe of my shoe on a small nub, shift all my weight briefly to that left toe, then swing my right foot over and across to the next proper ledge—all without any real handholds for balance.

My climbing partner, Maura, stood below me, holding the other end of the rope that secured me to the bolted metal anchors at the top of the climb. If I fell, she would pull down on the rope, using her belay device to stop me before I'd plummeted more than a foot or two. Climbing on top rope, as it's known, involves very little real risk. But my lungs constricted anyway, and I fought to squelch my dizziness and panic. From the ground, my friends encouraged me: "Trust your shoes!" "Trust your feet!" "This will be fine!" "You can do this!"

Finally, I took a deep breath, stepped forward, shifted my weight from one foot to the other, and made it across. I fumbled above my head for handholds to steady myself, then grinned and tried to breathe. For a moment, while I was in motion, I had felt weightless, in control. Unafraid. Now the fear came seeping back as I continued climbing, scrambling through the loose dirt that had collected on the ledges and lumps of rock in the crack. I finished the climb, but raggedly, fending off panic the whole way. It was a good start, but as Maura lowered me back to the ground, I knew I had a long way to go.

I didn't know any of the history when I decided to build myself a DIY exposure therapy program. Systematic, gradual exposure just seemed logical to me; I suppose maybe I'm a behaviorist at heart. But knowingly or not, I had opted for a program built on the work of Watson, Cover Jones, and Wolpe—and especially on one of Wolpe's proteges, the Israeli psychologist Edna Foa. Foa is now the director of the University of Pennsylvania's Center for the Treatment and Study of Anxiety. But as a postdoctoral fellow at Temple University in the early 1970s, Foa trained under Wolpe. Wolpe's work emphasized "imaginal" exposure; for instance, having an arachnophobic patient imagine a spider at a distance, and then imagine it slightly closer, and so on.

Foa's innovation was investigating whether a greater degree of "in vivo" exposure—exposure to the real fear stimulus, not just an imagined one—could improve on Wolpe's promising results. Earlier researchers had assumed such direct exposure could be dangerous for patients with phobias and anxiety disorders, but the science on that front was changing. Foa didn't go as far as some other clinicians had (in particular, a technique called "flooding" involved intense, even brutal, immersion), but she started to push harder within the system that Wolpe had developed. "I started to do studies of exposure in vivo, starting not with the highest level of fear but with moderate levels, and going faster, proceeding to higher and higher situations that evoke higher and higher anxiety," Foa told me. The results, she said, were "excellent."

As Mary Cover Jones's work with Little Peter suggested, exposure therapy is basically an inversion of classical conditioning. If you can teach an animal to expect pain from a blinking red light by repeatedly combining the light's appearance with an electrical shock until the animal reacts fearfully to the light alone, it makes sense that the twinning of stimulus and fear can be unraveled, too. Show the animal the red light enough times without an accompanying shock, and eventually it will no longer fear the light—a process known as extinction. Though it's worth noting that because our fear memories are designed, for survival reasons, to be sturdy and long-lasting, extinction can be a slower and weaker process than the initial conditioning. That's part of what makes the curing of fears so hard.

We don't know exactly what happens in the brain during the extinction process. As Foa put it to me, "Is it that you erase the connections" between stimulus and fear "or that you replace them with a new structure?" Her hypothesis is that exposure therapy trains the brain to create a second, competing structure alongside the fearful one. The new structure, she explained, "does not have the fear, and does not have the perception that the world is entirely dangerous and that oneself is entirely incompetent." When exposure therapy works, then, it's because the new structure has managed to override the old one.

That was why my panicked success in the Rock Gardens that day was really no success at all. I had climbed the wall, sure, but I had failed to convince my brain to begin building a new structure. Repeatedly terrorizing myself wouldn't solve anything; it wasn't enough to scramble through with wild eyes and a pounding heart. I had to learn to stay calm.

The rock was cold enough to numb my fingers. It was October 2, winter was reaching out for us, and I was on my eighth and final climbing excursion of the season. All summer, I had gone climbing every time someone with the necessary expertise and gear was willing to take me along. I had tried to systematize my outings, repeating the same routes to see if I could get farther, and stay calmer, each time.

In previous years, I would have pushed myself until my panic was unbearable, hoping that I could pop it like a soap bubble if only I tried hard enough. But now my strategy was to go only as far as I could without paralysis setting in. The goal was to build up the alternate structure in my brain that said, *This is OK. You are safe.* Then I'd come down before the old structure could reassert itself and hope to get a foot or two farther the next time around.

For this last outing, Ryan, Carrie, Maura, and I were at Copper Cliffs, a crag in Whitehorse's semi-industrial backyard, once a booming copper mining area, now a maze of quarries and mountain bike trails and small, shallow lakes. I was climbing Anna Banana, a short, beginner-friendly, sixteen-foot route up one side of an arête, a sharp wedge of rock protruding from the main cliff face. My first steps had been on easy footholds, gaps cutting into the leading point of the wedge, and I had no trouble until my feet were seven and a half, eight feet off the ground. I stalled out there, my right foot resting on a good ledge just around the corner of the arête while my left toe was tucked into a little cubbyhole a foot below. To continue, I had to pull my left leg up several feet to the next good hold.

I raised my arms and patted the rock above my head, blindly seeking out handholds I could use to pull myself up higher and give my left foot a fighting chance. I tend to rely on my hands and arms first, even though my legs are exponentially stronger; we're less accustomed to trusting a narrow toehold than a fist clamped around something solid. But I didn't find what I was looking for, so instead I spread my arms out wide and locked my fingers around the best stabilizing holds I could reach. Then I pushed off with all my weight on my right foot, pulled my arms tight to keep me close to the rock face, and scraped my left foot up the wall until it found the next hold, just as my right toe lost contact with the rock. I balanced there for a moment, raised my hands to solid holds now within my reach, and pulled up my dangling right foot.

I had done it. More importantly, I had done it calmly and coolly, without needing extra minutes to fight off panic, without groaning and moaning before I gave it a try. Maura lowered me down so I could climb up and do it again—more confidently, with even less hesitation. This time, I kept going, through a series of easy moves to the top of the route, where I reached up and smacked the anchor bolts in triumph: a touchdown spike. I did a quick mental survey of my body. My breathing was steady, my head clear. For one day at least, I had successfully redirected my brain to reject fear.

Compared to living with PTSD or broader anxiety disorders, or some harder-to-navigate phobias, my fear of heights is trivial. It doesn't keep me awake at night, or ruin my relationships, or bleed into every area of my life. If I moved back to the flatlands and avoided high-rise balconies, dodging my symptoms by practicing avoidance, I would hardly notice it.

Still, it can limit me. I would have liked to climb that mast high into the rigging during my too-brief sailing career, or enjoy the view over Florence. Sometimes I get scared on steep stairways or balconies with flimsy-seeming railings, and I have still never climbed a tree. Taken individually those are all tiny things, but they add up to a feeling of helplessness: My choices are not entirely my own.

During the winter that followed my exposure therapy experiment, I kept climbing: at big indoor gyms in San Francisco and Vancouver and on small, homemade climbing walls at home in Whitehorse, in local schools, and a friend's basement. By my standards, I made substantial progress. Gradually, I found I could climb higher—six, eight, ten, twelve feet—before my chest started to constrict and my pulse started to pound in my ears. Sometimes I could complete an entire short route without feeling afraid at all.

But even as I improved, I felt my priorities changing. I still didn't actually *enjoy* climbing, really. It was like medicine, something slightly unpleasant I consumed because I thought that in the long run, it would do me good. I started to wonder if this particular self-prescribed treatment had given me all the benefit I was going to get from it, if it might make more sense to spend my time doing things that made me happy. Was I still practicing avoidance if I chose joy over stress and struggle? I wasn't sure.

6

wrecked

When I first began this project, and began to map out the various ways in which I felt afraid, I thought of the specific phobias as a kind of pure-blooded example of raw fear, uncomplicated, straightforward: See a spider, be afraid; Phobos bringing fear to the battlefield, not Deimos bringing dread. Other strains might be murkier, but a phobia, I thought, was fear undiluted.

But as I learned more, and reflected further on my own experiences, I realized that things weren't quite so simple. Yes, my fear of heights had a specific, identifiable trigger: exposure to heights from which I felt I could fall. In theory, heights provided the clear and present threat to my safety that the definition of fear (as opposed to anxiety) requires. But as we've seen, the triggers for my fear weren't always truly threatening, and my response was often far from reasonable.

I'd never thought of myself as a particularly anxious person, in general, let alone to a clinical degree, but when I told the story of what happened that day on the Usual to a friend who has wrestled with anxiety issues for years, she stared at me as though I was missing something obvious. "Eva," she said, "that was a panic attack."

She was right. And not just about that incident, either, but probably about all the times I had been so terrorized by my fear of falling that I had frozen in place or curled up on the ground and refused to move, unable to breathe—incidents I had always referred to as "freak-outs" or "meltdowns" if I referred to them at all. It was as though my fear of heights was an anxiety disorder in miniature—anxiety focused through a narrow lens. There was no clear, bright line between the two.

I understood those "freak-outs," those eruptions at the intersection of fear and anxiety, better now, even if my homespun exposure therapy program hadn't necessarily relieved me of them. And I remained in a more peaceful place than I'd previously occupied in relation to my fear of death and loss. Now it was time to face the third major strain of fear I had planned to tackle and see if I could find some relief there, too.

If the line between fear and anxiety is murky, trauma blurs the boundaries even further between a person's perception of specific, immediate threats and their more generalized state of anxiety. Post-traumatic stress disorder is another unhappy cousin in the extended family of phobias and anxiety disorders. If fear is "anticipatory pain," then you could say that trauma is a state of anticipatory pain driven by pain from your past. It's the refusal of past fearful memories to let you go.

Most of us associate PTSD with dramatic and unusual threats to our lives: war, most often, or an earthquake, or a gun held to your head at the cash machine. But trauma can be much more quotidian and unremarkable, too. "Trauma," broadly defined, can refer to any deeply distressing or disturbing experience, whether that means a physical injury, a frightening event, or even witnessing something bad happening to someone else. Traumas can be one-offs, the scary moments we live through and then leave behind, but they can also linger in our systems, causing serious long-term health problems.

And while the vast majority of traumatic events never develop into full-blown PTSD, many of us carry traumatic memories around with us every day—memories that have the potential to insert themselves into our minds at the worst moments, hijacking our ability to determine threat from safety. Is the thing that's scaring us imminent and concrete, or is our fear abstract and irrational? When trauma is involved, nothing seems clear anymore.

As with anxiety, I had never thought of trauma as being a part of my life. After all, nothing *really bad* has ever happened to me. I'm a lucky person: I've had a good life. Right?

But I was wrong again.

SUMMER 2000
Somewhere in Eastern Ontario

I don't remember the exact date or the exact road anymore. I know that my friend Erika and I were savoring the summer before our last year of high school, staying at my stepdad's cottage for a couple of unsupervised days. I know that we had driven my old Chevrolet—a 1987 Celebrity with two wide bench seats and a trunk big enough to live in—into the nearest town for ice cream. I remember that I had a mix tape in the Chevy's tape deck, the tinny music cranked.

I was driving a little too fast on a rough gravel road. Not outrageously so, but I didn't realize that a highway crew had laid down fresh gravel only the day before, and as I came over a small, sharp hill, I saw the curve ahead too late.

I don't remember braking, although my dad said later that I must have jammed my foot down on the pedal. What I remember is the sudden alarm when I saw the curve, the sharp twitch I gave the steering wheel, the floating feeling of my tires losing traction on the shifting gravel, and the slow but accelerating pendulum swings of the big old car as it began to fishtail.

I did everything wrong. I kept cranking on the wheel, trying to catch up to the fishtail and recover control, but instead each swing got wider and wilder, until it felt like we were on an amusement park ride—the kind designed for maximum puking. We careened down the road like that for a hundred, maybe two hundred, yards, and then the horizon whirled around us and we were upside down in the ditch, wheels still spinning, Jennifer Lopez's "Feelin' So Good" still squeezing out of the stereo.

As soon as the car had stopped moving, Erika unbuckled her seat belt and rolled out through the passenger-side window, but I took a while longer to react. I remember hanging upside down by my seat belt, dangling, staring at the windshield's new map of cracks.

I was strangely calm, hanging there. For a few moments, nothing existed except the splintered windshield that filled my vision. It was a little like the time I'd been knocked unconscious in a hockey game, earlier that same year: I'd opened my eyes as I lay facedown and stared and stared at the tiny crystalline structures just visible below the surface of the ice. I was completely absorbed in those tiny formations for what felt like several long minutes before I remembered who I was, where I was, and that I should try to stand up.

In the car, I finally came back to myself and tore my eyes away from the ruined windshield. Erika was crouched in the ditch beside my window, asking me if I was OK, her concern climbing the longer I remained silent. I slowly reached out and unrolled the window the rest of the way. I unbuckled my seat belt and fell onto the inside of the car's roof, then crawled out the open window, cutting my knees on broken glass as I went.

We were extraordinarily lucky. Erika had bumped her head on the roof as we went over; I had one bad cut on my left kneecap, where I still have a faint scar. The police officers who responded to the crash drove us to a neighbor's cottage so we could call our parents, and a flatbed tow truck hauled the wreckage of the Chevy away. (The front driver's-side tire, the one I assumed had taken the main force of the roll, was ripped clean in half. Still in shock, I suggested repeatedly to the cops that the car just needed a new windshield and a few dents punched out of the roof.)

Later, I paid a few hundred dollars out of my savings to a local wrecker who'd stored the car after the crash. My dad had been away on vacation when it all happened, and it took a few days to sort things out, and in the meantime they had the meter running, racking up my bill. But that was the only real consequence I faced.

Soon I was back on the road, in another enormous Chevy—this one a 1989 Caprice Classic, an even bigger beast than the one I'd slain. Erika's parents forgave me, I think, and then I went off to university and didn't do much driving for a few years. A couple of times, in my early twenties, I found myself on gravel roads, maybe going around a curve or climbing a steep hill, and the feelings and memories would come back in a heady rush: those heavy, wild pendulum swings; the fear and confusion as I cranked the wheel; the cold, calm moments of communion with the windshield after the crash.

By the time I'd finished university, the flashbacks had gone away. This, I would learn years later, is a pretty commonplace example of a traumatic event that heals itself with a bit of time.

JUNE 11, 2014

Near Northway Junction, on the Alaska Highway

This is the one I remember most clearly. Every detail is chiseled into my memory: the view of the Wrangell Mountains serrating the horizon in the distance, the U-Haul truck barreling down the hill and around the bend toward me, the truck's front driver's-side tire sliding across the yellow center line by a few inches at first, and then more, and then more.

I remember wondering when the other driver would notice his drift into my lane and correct himself. I remember looking up at him and seeing that his gaze was on those distant mountains, not on me or the road at all. I remember realizing I had no time to wait and hope, no time even to honk. I remember hauling my steering wheel as hard as I could to the right, flattening the brake pedal, tensing my body for impact, and closing my eyes, believing that I would never open them again. I remember thinking, *This is it.*

When I did open my eyes, my Jeep was stopped on the narrow

shoulder and the U-Haul was receding in my rearview mirror. My side mirror was gone, my windshield shattered, and as I checked myself for injury, dumbly patting at my chest and head and thighs, I realized I had avoided the head-on collision I'd expected. I had escaped with a sideswipe, had escaped by inches the force of my engine block bursting through the dashboard. I was not riddled with metal and bleeding out. Images of the short future I might have had flooded through my mind: the Jaws of Life cutting me free, the helicopter ride to a Fairbanks ICU, the tubes and the blood bags and my parents picking up the phone a continent away.

In my rearview the U-Haul was still visible, slowing to a halt a ways down the highway. I felt an enormous surge of energy, unbuckled my seat belt, pushed open the flayed remnants of my door, and jumped out onto the pavement. I've always thought of myself as more prone to flight than fight, but now I ran down the highway toward the distant truck, shaking my fists in the air, hollering. By the time I reached the U-Haul, the driver was emerging, an older man, shaken, trying to apologize. I pushed my face close to his, still yelling.

"What the hell, man!"

"I'm sorry . . . I'm sorry," he tried to say. "Are you OK?"

"Yeah, but my car is *fucked up*!"

Slowly my rage and adrenalin ebbed away. He turned his truck around—it was undamaged apart from the big driver's-side mirror, which had exploded my windshield—and since we were far from any cell service, we convoyed north, to the small Alaska town of Tok, to report the crash to state troopers. I checked into a motel there, and the man and his family bought me dinner before they carried on south again. ("You don't have to do that," I said, automatically, when he picked up the check. "I really do," he said, and we both managed to laugh.)

After dinner, I decided I needed a shower. When I took my clothes off, tiny shards of glass sifted onto the motel room carpet. I was covered in them, I realized: Shards from the windshield were in my hair, embedded in my T-shirt, shimmering on my collarbone like glitter. The soft skin on the inside of my right forearm and the tanned, tougher skin on the outside of my left forearm were pocked with tiny cuts—a pattern that had been formed, I figured out, when I had the wheel cranked hard over to the right. Suddenly, all I wanted in the world was to talk to my mom. I called both my parents, put up a post on Facebook urging my friends to hug their loved ones, and went to bed.

The strangest thing about a near miss with death is how life just carries on afterward. There is no ceremony or ritual to mark the enormity of the moment, and most people around you may not even realize what has happened. Without any real injuries, I felt silly even dwelling on the almost and the not quite. But I couldn't shake the memory of how certain I'd been, when I closed my eyes before impact, that I was experiencing my last moments of consciousness. *This is it,* I'd thought. It wasn't the first time I'd experienced the fear of dying, but I'd never been sure of my imminent death before. Being proved wrong, but only barely, left me wrestling with a confusing mixture of gratitude and horror.

In the memoir *I Am, I Am, I Am,* British author Maggie O'Farrell describes seventeen different times in her life in which she confronted her own possible death.

> There is nothing unique or special in a near-death experience. They are not rare; everyone, I would venture, has had them, at one time or another, perhaps without even realizing it. The brush of a van too close to your bicycle, the tired medic who realizes that a dosage ought to be checked one final time, the driver who has drunk too much and is reluctantly persuaded to relinquish the car keys, the train missed after sleeping through an alarm, the airplane not caught, the virus never inhaled, the assailant never encountered, the path not taken. We are, all of us, wandering about in a state of oblivion, borrowing our time, seizing our days, escaping our fates, slipping through loopholes, unaware of when the axe may fall.

She goes on, "If you are aware of these moments, they will alter you. You can try to forget them, to turn away from them, to shrug them off, but they will have infiltrated you, whether you like it or not. They will take up residence inside you and become part of who you are, like a heart stent or a pin that holds together a broken bone."

I did feel altered. I felt like I had glimpsed something I shouldn't have—something secret, something almost indecent. It was as though, for a moment, I had gained some supernatural ability to see my own alternate future, and it was an awful one. I couldn't stop thinking about that path not taken, those images that had filled my mind after the crash: the Jaws of Life, the helicopter, and the tubes and the blood bags. Most of all, I thought about my parents. Death, when it's quick, is mostly something that happens to the people left behind. I was grateful, for their sakes, that I hadn't become a riddle of shrapnel and blood.

Later that summer, I spent two weeks roaming the rugged backroads of northern British Columbia in a comically tiny rental car. It was a long-planned work trip; I had intended to drive down in my Jeep, but the Jeep was no more. So instead I drafted in the wake of enormous logging trucks in a Fiat 500, driving through torrential rain and thick clouds of wildfire smoke.

I was still, at that point, a functioning, competent driver. My fear had not yet compromised me. But I realized, on that trip, that the collision with the U-Haul had eroded a critical element of driving: my trust in other drivers. The whole motor-vehicle system is predicated on our reliance on others: to stop at red lights, to signal before turning, to *stay in their own damn lane*. Where I had once been able to go on faith, to blithely trust other drivers to do their part, I was now doubtful, suspicious. I didn't know anything about trauma, how it can lead to guardedness and vigilance verging on paranoia, how it can make people wrongly believe, as clinicians often put it, that "the world is entirely dangerous."

I didn't know those things, but now I was living them. I kept the Fiat edged over almost onto the shoulder, as far from the yellow center line as I could get. I watched the tires of oncoming cars, to see if they started to drift my way. I was nervous. I was scared.

JANUARY 8, 2016

outskirts of Whitehorse, on the Alaska Highway

January 1 was supposed to be the day I made my new start after my mom's death. I had set the date for myself months earlier, drawn to the easy symbolism of the new year. It would be five months and one week since we'd watched the hospital staff turn off the machines that were keeping my mother's body alive, and just one week since I'd finally come home from more than two months on the road, driving from the Yukon to California and back in a desperate effort to stay in motion through my grief. Now that I was back in my apartment, I'd decided, I would put the days-long Netflix binges and the endless platefuls of delivery chow mein behind me. I had stagnated long enough. It was time to put myself back together.

On New Year's Day, I packed my aging Toyota 4Runner, the car I'd bought to replace the Jeep after its encounter with the U-Haul, and headed north. I wound my way up the North Klondike Highway to sleepy Dawson City, and then up the snow-covered Dempster Highway—the only Canadian road that crosses the Arctic Circle—to Inuvik, on the Mackenzie River delta. Inuvik was, at that time, the end of the permanent road network; from there, I piloted my car down a steep, snowy hill onto the frozen surface of the great river itself, and followed a wide ice road ploughed along the Mackenzie and out to the village of Tuktoyaktuk, on the shore of the Beaufort Sea.

It was the first real work trip I had attempted since the cruise in August. I was planning a potential book about the past and future of the fabled Northwest Passage, the legendary sea route through the Canadian Arctic. My idea was to see the Passage's outlet, the Beaufort Sea, in its full, frozen winter glory, to imagine myself locked in the ice and the darkness of the old wooden ships, and to visit a community that stood to be irrevocably changed by the development of an Arctic sea passage.

As the days passed, I felt good; I felt like my old self, almost. I looked out at the world around me, and I was curious and engaged again. I walked on the frozen shore of the Beaufort in the relentless blackness of an Arctic winter morning, and I could almost feel the presence of the scurvy-addled sailors trapped in their ships. I had bought myself a brightly colored pair of trail runners on sale after Christmas, hoping that I could fashion myself into the sort of person who ran their woes away, and I jogged along the ice road in the afternoon twilight, iPod pounding in my ears, Ski-Doos buzzing by. I was beginning to really believe I would not remain a sad husk of a person indefinitely. I was going to be OK.

I made the drive back south in short, easy stages—no more than five or six hours behind the wheel each day. As I eased the 4Runner down the Dempster's narrow curves, watching the mountains march away on either side of me, trying not to look at the places where the road was skirted by nothing but a long cliff, I thought back to the accident in high school. A decade and a half later, I could still remember the sickening power of the big old Chevy as it swung back and forth on the gravel, fishtailing in ever-widening arcs before it heaved itself over into the ditch. I could remember how it felt to hang upside down by my seat belt, staring at the mad spiderweb of smashed windshield, the tape deck still squeezing out that Jennifer Lopez song.

As I drove along the Dempster, I thought, *I don't ever want to feel those sensations again.*

On January 8, I set out from Dawson City on the final leg for home. An hour outside Whitehorse, I disengaged my four-wheel drive—the pavement seemed clean and dry. Eight miles from my downtown apartment, I turned onto the familiar snow-edged blacktop of the two-lane Alaska Highway and found myself puttering along behind an old pickup truck, doing forty-three miles per hour in a fifty-five zone.

After eight days of rigorously patient, cautious driving, I'd had enough. I wanted to get home. I wanted to change my clothes, shower, eat a meal that didn't involve overpriced crinkle-cut french fries. Drink a good cup of coffee with real cream instead of the powdered whitener that sits in industrial-size shakers on restaurant tables and coffee counters across the North American Arctic. I checked for oncoming traffic, turned my wheel to pull out and pass, pushed down on the gas pedal—and felt the thing I'd been dreading for a week. My tires were losing traction on unseen ice. Instead of smoothly changing lanes, the SUV began to spin—once around, maybe twice, it was hard to tell, the highway a white blur around me—and careened across the opposing lane toward a tall snowbank. *I'll hit the bank and stop*, I remember thinking, weirdly calm as I sat powerless in the driver's seat.

It's easy to forget, when your car is running normally, under control, about the magnitude of the forces involved in sending two tons of metal hurtling down a highway. My spinning Toyota hit the snowbank broadside, slid over and through it, and rolled—once? twice?—down a small hill into the ditch. I could hear more than I could see: the fireworks of windows shattering, loud thumps as the car's body made repeated impact with the frozen ground, and the smaller bumps and clatters of my belongings flying around inside the vehicle.

When the 4Runner stopped moving, it lay on its side, the driver's door flat on the ground and the passenger door in the air, where the roof should have been. I reached up to my right hip and unbuckled my seat belt, then stood carefully inside the vehicle, my boots crunching on the shards of my broken driver's-side window. The windshield had vanished entirely, ripped out in one whole piece. I ducked down and walked out through the gap where the glass should have been.

Outside, half a dozen cars had pulled over and several shadowy figures were hurrying down into the ditch toward me. My glasses were coated with a spray of snow and ice. I pulled them off and touched the top of my head, where I could feel slivers of glass, grit, and a large lump already rising.

My brain was sluggish. I answered the bystanders, and later the paramedics, haltingly; forming clear sentences felt like lifting weights. I managed to ask that my purse, my cell phone, and my laptop be retrieved from the 4Runner before the ambulance took me to the hospital. As we drove away, a young EMT checked my blood pressure, my pupils, strapped me into a neck brace as a precaution. But she didn't seem worried. And once the shock and the adrenalin wore off, I was indeed fine—nothing but a goose egg on my head to show for a Hollywood-caliber crash.

Still, the rollover felt like a personal rebuke. I had just been getting back on my feet again, and it was as though the universe had said, "Oh, no, not yet," and knocked me back down. The timing felt cruel. I had been so careful! So cautious, up until that last hour, those last minutes. I had known that the road trip would be potentially dangerous, but to make it across the Arctic Circle, all the way to the frozen ocean and back again, before crashing in the Whitehorse suburbs? It didn't seem fair. I remember wondering, *When is my luck going to change?*

APRIL 30, 2016

South of Fort Nelson, on the Alaska Highway

For three months after the 4Runner crash, I didn't own a vehicle. I walked, I borrowed a friend's car for a stretch, I rode the bus to my physiotherapy appointments. My neck, as it turned out, had been strained in the crash, and it took weeks of work to untangle the knots in my muscles and nerves. I was just as happy to wait out the winter before driving again; I had a suspicion that the latest crash might have damaged more than just the nerves in my neck.

I didn't bother looking for a new car, because there was one waiting for me in Arizona. My mom and stepdad had spent their winters there for the past few years, in a bungalow on the outskirts of Phoenix. Back in the fall, after my mom died, my stepdad had driven her beloved red Subaru hatchback down there so that guests could use it when they came to visit; he couldn't bear to sell it, and he'd offered it to me, but at the time I already had a car, so south it went.

I'd emailed him while still in a hospital bed in Whitehorse, after the rollover, to see if his original offer still stood. In early April, I flew south for a visit and to collect my new wheels.

I took my time driving north again, moseying from Moab to Boise, Seattle to Whistler. The driving, through hot, dry country, went fine. But by the time I was heading for northern British Columbia, late in the month, a grim premonition had set in. Originally I had thought about putting the car on the ferry from Washington State to Alaska, to save three days of the most remote driving of the trip. But with an unfavorable exchange rate, the ferry ticket came to almost 1,500 Canadian dollars, so I had set the idea aside.

Still, leaving Whistler, I almost turned south for the border again, almost said screw it and put the car on the ferry and the charges on my credit card. I had a bad feeling about the drive, and I couldn't shake it.

A day and a half later, in the early evening, I was closing in on Fort Nelson, my destination for the night. From there, I would have one full day of driving left, and then I'd be home. I had nearly stopped a few hours earlier, in Fort St. John; I'd still been unable to escape my grim thoughts and wondered if I should get off the road. But I'd decided to press on. I wanted to get home.

Weather can change fast in the mountains. I was driving through hilly country on the northernmost fringe of the Rockies, not too far from where I'd had my breakdown on the ice-climbing trip a couple of months earlier. Sudden, short-lived downpours of rain fell, on and off, as I drove, as though someone were fiddling with a tap. And then I came over the crest of a hill and saw something on the road surface below me, at the bottom.

I wasn't speeding, but I slowed a bit further and squinted through my windshield. Why did the road up ahead look so strange?

Moments later, I hit the hail. It was everywhere, freshly fallen, covering the pavement thickly for maybe two hundred yards before it ended as abruptly as it had begun. It was like ball bearings: I might as well have been back on that newly laid gravel again, in high school. I felt my tires lose contact with the road—that now-familiar feeling.

But I'd grown and learned since that first crash, and I stayed calm at first and steered the slowing car straight down the center of the highway. Time stretched out. I gripped the wheel and stared at the road, willing the damn machine to point straight, and felt fear creeping up through my chest. In the Jeep, there'd been no time to feel afraid, and in the 4Runner, I hadn't thought I had any reason to be afraid.

When the Subaru started to fishtail, gently at first, the fear burst out of me and I screamed into my windshield, "No, no, not again!" The car swung left and cruised nose-first into the ditch, crunching into the far side of the little valley. Then, almost as an afterthought, when I thought the worst was over, it flipped over onto its roof. I was hanging upside down by my seat belt again.

Again, I crawled through a driver's-side window on hands and knees—this time into a ditch filled with slushy water. Again, I searched through an upside-down vehicle for my phone and my purse. Again, I was eventually surrounded by bystanders; the first to arrive fishtailed through the hail and nearly went off the road, too, even though my car in the ditch served as a warning to slow way down.

A group of seasonal workers on their way to Alaska wrapped me in a blanket, gave me a granola bar, a bottle of water, some dry socks. "You seem really calm," one of them said, and I answered, "I know the drill," not bothering to explain. One local woman drove ahead, into cell range, to phone the police. A young family with room to spare in their truck loaded up my belongings, my camping gear, and me. Since I'd crawled out of the car, I'd been almost robotically level—feeling flat, speaking in a monotone, existing in a numb state of disbelief—but when the young mother leaned forward from the back seat of the truck's cab and said, "Someone was watching over you," my calm dissolved. I couldn't breathe. My eyes were full and my chest was tight, my throat burning. I was afraid that if I let myself start to cry, I wouldn't stop for a long, long time.

The nurses at the Fort Nelson hospital kept me overnight for observation, although mostly, I think, they knew I had nowhere else to go. They put me in an empty room in the children's ward, and I curled up on the small, plastic-covered mattress, with full-size painted murals of Pocahontas and a few of the 101 Dalmatians looming over me, and stared into the darkness until I finally fell asleep. I felt awful. It wasn't just that I had crashed again (what was *wrong* with me?) but that I had destroyed the little red Subaru in the process. A friend who'd heard about the crash texted me right away to say she understood. The first time she'd broken something that her dad had given her before he died, she'd been a wreck. It was a hard thing to describe to someone who hasn't been there, but it felt like I had betrayed my mother somehow. She was gone, and I was still letting her down.

The next day I rode the Greyhound bus home, feeling fragile and embarrassed. I went sheepishly back to my physiotherapist that same week; my neck needed to be unknotted once again. But it didn't take me long to realize there was more to the damage this time. A few days after the accident, I drove a rental car through a rainstorm and had to pull over on the side of the highway to sob and catch my breath. Every time I went around a sharp curve on the wet road, I imagined myself rolling over and over into the ditch. I replayed the feeling of the tires losing traction, pictured it becoming reality again, and I panicked.

The problem was the same a few weeks later, when I drove

to the launch point for a canoe trip in a newly acquired used car. I was OK if the pavement was dry, and straight, and flat, but curves and hills, and any amount of water on the road, sent my mind reeling into the ditch. I could imagine—not just imagine; I knew!—how it would sound, how it would feel.

When winter came, things only got worse. I became so cautious, I was a hazard. I drove so slowly on snow and ice that other drivers tore around me, no doubt cursing as they passed. Friends joked that, after all my crashes, I must be invincible. But I felt certain I had used up all my chances. I was fixated on the possibility of another crash, and certain that I wouldn't survive it.

Trying to comfort me, my friend Eric pointed out that driving is the least safe thing most people do on any given day but that most of us are comfortably oblivious to its risks. I'd simply had my illusion of safety stripped away, he said, and now the risks felt real and close all the time.

It wasn't just that, though. The U-Haul sliding across that center line had eroded my trust in other drivers, but the two rollovers, coming one right after the other, had radically undermined my trust in myself.

It took me a long time to accept that I had a right to use the word—after all, I had walked away from each crash largely unharmed, right?—but there I was, nonetheless. *Traumatized.*

In 1872, Charles Darwin published *The Expression of the Emotions in Man and Animals*, a follow-up to *On the Origin of Species* and *The Descent of Man*. In it, he described the physical ways that humans display their escalating levels of fear. When they reach an "agony of terror," he writes, "all the muscles of the body may become rigid, or may be thrown into convulsive movements. The hands are alternatively clenched and opened, often with a twitching movement. The arms may be protruded, as if to avert some dreadful danger, or may be thrown wildly over the head. . . . In other cases there is a sudden and uncontrollable tendency to headlong flight; and so strong is this, that the boldest soldiers may be seized with a sudden panic."

There was something familiar in that description, something that hinted at a condition still a century away from widespread clinical attention and recognition. During the muddy slaughter of the First World War, the Brits called it "shell shock." I remember seeing the jerky black-and-white footage in a high school history class: young men twitching madly, eyes wild, apparently driven to a breakdown of control over their bodies by the relentless horror of the trenches. Initially these soldiers were eligible for treatment and pensions, but as the diagnosis spread further and further, the alarmed military brass tried to suppress the problem.

"Caught between taking the suffering of their soldiers seriously and pursuing victory over the Germans," Bessel van der Kolk writes in *The Body Keeps the Score*, "the British General Staff issued General Routine Order Number 2384 in June of 1917, which stated, 'in no circumstances whatever will the expression "shell shock" be used verbally or be recorded in any regimental or other casualty report, or any hospital or other medical document.' All soldiers with psychiatric problems were to be given a single poker-faced diagnosis of 'NYDN' (Not Yet Diagnosed, Nervous)." The understatement would have been funny if it weren't so damaging.

Medical research into how to treat the condition was discouraged or suppressed, too, until a renewed outbreak, in the Second World War. And this pattern repeated itself again, in the United States at least, with the Vietnam War. When van der Kolk, then a young psychiatrist, attempted to treat Vietnam veterans struggling after their return from Southeast Asia, he was dismayed to discover that there were no resources specific to the condition available to him—no textbooks, no protocols.

He writes,

> In those early days at the VA, we labeled our veterans with all sorts of diagnoses—alcoholism, substance abuse, depression, mood disorder, even schizophrenia—and we tried every treatment in our textbooks. But for all our efforts it became clear that we were actually accomplishing very little. The powerful drugs we prescribed often left the men in such a fog that they could barely function. When we encouraged them to talk about the precise details of a traumatic event, we often inadvertently triggered a full-blown flashback, rather than helping them resolve the issue. Many of them dropped out of treatment because we were not only failing to help but also sometimes making things worse.

Edna Foa, the Joseph Wolpe disciple who'd taken the lead on intensifying and formalizing exposure therapy for phobias and anxiety, remembered the same void. "We didn't have any studies on PTSD," she told me.

In 1980, PTSD was included for the first time in the *Diagnostic and Statistical Manual of Mental Disorders*. The inclusion at least formalized the problem. Since then, trauma therapy has undergone a revolution, driven in part by an influx of military veterans from the Gulf War and the seemingly endless post-9/11 wars in Iraq and Afghanistan. And as clinicians' understanding of PTSD has grown, so has our awareness of its staggering reach. We now know that PTSD affects not just soldiers and civilians emerging from war but also drone operators who've never left their home base; first responders, from beat cops to search-and-rescue volunteers operating out of luxurious mountain resorts; survivors of car wrecks, assaults, and less obvious forms of trauma. An estimated eight million Americans experience PTSD every year.

During the First World War, the Germans treated their shell-shocked soldiers with electroshock therapy. In the Second World War, hypnosis became a popular option. Over the decades since, treatments for PTSD have often followed the same broad trends as other anxiety-related disorders; there has frequently been overlap between the ways phobias and PTSD are treated.

For instance, exposure therapy, the treatment I'd tried to DIY in an effort to ease my fear of heights. Back in the early 1980s, when the diagnosis was still new, Edna Foa wondered whether the exposure therapy treatment she had developed for phobias and obsessive-compulsive disorder might be workable for PTSD, too. "I thought, well, this is an anxiety disorder, there is no reason why we cannot adapt the treatment, the exposure therapy treatment, to PTSD," she told me.

You can't reexpose someone to rape or a bomb, so Foa settled on a program of imaginal exposure for the traumatic memory itself, and in vivo, or actual, exposure to the secondary effects:

the patient's avoidance behaviors, which can perpetuate trauma's power. The imaginal exposure would be conducted in sessions with therapists, and the in vivo exposure came as homework: going to places that reminded the patients of the trauma, or to safe places they perceived as dangerous. That might mean walking a downtown street at night after experiencing a violent assault, or going to malls again after a mass shooting. Avoidance often means perpetuation, and so the long-held idea of "facing our fears" remains at the core of many of our treatment options.

Throughout the 1990s, Foa's team taught other groups of therapists how to administer what Foa called prolonged exposure (PE) therapy and how to monitor the results. They found that PE was effective in almost 80 percent of patients: between 40 and 50 percent became essentially symptom-free, while 20 to 30 percent still had some recurring symptoms but were much improved. "We're not 100 percent successful," she said, "but no treatment is."

Meanwhile, while Foa was working in Philadelphia to adapt her phobia treatment for trauma patients, another brand-new PTSD treatment—eye movement desensitization and reprocessing (EMDR)—was being developed in Northern California. This was the treatment I would wind up turning to in an effort to end my intrusive, fearful memories and panics while driving.

The idea of EMDR, in the simplest, pared-down terms, is that the patient moves their eyes back and forth in a rhythm while working through their traumatic memories with a trained therapist, and something about the movement—the actual mechanism isn't fully understood—helps to process them. It's as though the memories haven't been filed correctly, and so they stick out of the filing cabinet drawer in our minds, jamming the drawer open, intruding on our lives. In theory, EMDR tidies them away where they can't hurt us anymore.

Sounds pretty sci-fi, right? For several years after its invention, or discovery, plenty of scientists were skeptical, too. But over the past three decades, EMDR has proven effective in clinical trial after clinical trial. I was game to give it a try.

The treatment was born in 1987 in Los Gatos, California. For eight years, ever since a cancer diagnosis, Francine Shapiro had been seeking to understand the connections between our bodies and our brains, in sickness and in health. "My search took me cross-country," Shapiro writes in *EMDR: The Breakthrough Therapy for Overcoming Anxiety, Stress, and Trauma*, "and into dozens of workshops, seminars, and training programs. As the journey unfolded, it brought me into contact with myriad forms of psychotherapy."

It was a sunny spring day, and Shapiro took a break to stretch her legs. She left her office and went for a walk by a small lake. Ducks paddled in the water, picnickers lounged on the grass. As she walked, her brain went over and over something, worrying at it. Writing years later, she no longer remembered what it was, exactly, just that it was "one of those nagging negative thoughts that the mind keeps chewing over (without digesting)." You know the ones: a memory of a time you were wronged that you can't seem to let go of, or, even worse, of a time you wronged someone else. One of those weird, tenacious thoughts or scraps of memory that can keep you up at night or ruin your day, refusing to be banished.

Suddenly she realized that the negative thought had disappeared from her mind. She summoned it back again, but now, she found, it had lost its negative power. It felt neutral.

Shapiro, attuned to look for connections after her years of research, tried to reconstruct what had been happening in her body when her mind sapped the thought of its powers. She realized she had been working her eyes back and forth, rhythmically, as she walked and chewed on the thought. Intrigued by the possibility of a link, she tried it again. She summoned up another nagging, anxious, negative thought, and then she rapidly flicked her eyes back and forth.

"That thought went away, too," she wrote. "And when I brought it back, its negative emotional charge was gone."

Over the following months, Shapiro continued practicing and experimenting on herself, on friends, and eventually on seventy volunteers. She had people tell her about their troubling memories while she led them through sets of eye movements. She refined her methods, wagging two fingers on a regular beat twelve inches in front of a subject's eyes to induce the back-and-forth. She would ask subjects how they felt after each set and then ask them to concentrate on that feeling during the next round, going deeper and deeper.

In late 1987, Shapiro designed her first formal study of the treatment. One group of traumatized subjects received her nascent EMDR method: the guided eye movements. To create a control group, she treated a second batch of subjects with all the verbal components of the EMDR treatment: the subject telling the story, Shapiro's follow-up questions about their feelings, the requests for them to concentrate on those feelings. She kept everything the same except for the eye movements, to see if she could isolate their effect. ("Unfortunately," she noted in her book, "talk therapy has been shown to have about as much success in treating PTSD as a sugar pill." So using it as a placebo for the controls seemed appropriate.)

Shapiro published the results of her study in 1989 in the *Journal of Traumatic Stress* and the *Journal of Behavior Therapy and Experimental Psychiatry*. She found that a single session of EMDR had decreased or eliminated the most intrusive symptoms—flashbacks, nightmares, panic—in 100 percent of her subjects. In her article, Shapiro wrote that "the evidence clearly indicates that a single session of the EMD procedure is effective in desensitizing memories of traumatic incidents and changing the subjects' cognitive assessments of their individual situations." In other words, it worked, and it worked well. The effects, Shapiro found, were stable at a three-month check-in. EMDR had arrived.

But acceptance of the new method did not come right away. In 1993, the *Journal of Traumatic Stress*—home to Shapiro's original study—published a commentary by two clinicians who had undergone Shapiro's training, used it to little or no effect with their own clients, and were broadly critical of the new treatment procedure. The letter writers were particularly concerned by the apparent brevity of EMDR; they worried about a lack of connection between therapist and patient, and about people opting out of the hard work of therapy and instead falling for an easy fix. "We recommend extreme caution in embracing, using, or endorsing EMDR. . . . The atmosphere of hype and rush to embrace it as a panacea is disturbing," the authors wrote.

Psychiatrist and trauma specialist Bessel van der Kolk was an eventual and enthusiastic adopter of EMDR, but like many others, he was skeptical at first. "At that time I'd heard only that EMDR was a new fad in which therapists wiggled their fingers in front of patients' eyes," he wrote, twenty years after he adopted EMDR in his own clinical practice. "To me and my academic colleagues, it sounded like yet another of the crazes that have always plagued psychiatry."

In 1999, the *American Journal of Psychology* published a scathing review of a book on EMDR written by a clinician who'd adopted Shapiro's method. Under the title "The Power of Placebos," reviewer Bruce Bridgeman wrote that EMDR had "all the characteristics of a cult cure. It makes miraculous claims, backed by thin evidence; it is simple, yet can be learned only at the founder's seminars or in a handful of approved programs; it is based on some unique and idiosyncratic assumptions about the structure of the mind." Bridgeman greeted the book's assertions with words like "remarkable" and "amazing," dripping sarcasm. He acknowledged that Shapiro's original study was, "as clinical research goes, better than most." But he called for more trials with larger sample sizes. He would get his wish. Through the 1990s and into the 2000s, the clinical trials and peer-reviewed papers piled up. Today, EMDR is a widely accepted clinical practice.

But one big question remained unanswered: EMDR worked, it seemed clear, but how? No one was sure of the actual physical mechanism, or mechanisms, in the brain that made the treatment effective. It's a mystery that researchers are still working to solve.

In June 2018, I walked into a therapist's office in downtown Whitehorse. I remember that it was sunny; on my drive downtown, the pavement had been dry, and my body free of fear. I hoped that EMDR could teach me how to keep it that way.

I'd never been the "therapy type," I'd always thought. As far as I could tell, therapy was for rich people, or for deeply troubled people, or for New Yorkers in Hollywood rom-coms. My parents had tried sending me to a child psychologist after they split up, but I don't remember feeling anything but boredom in her office, boredom and a mild disdain for her repeated insistence that the divorce wasn't my fault. I had never worried that it might be. The highlight of my interactions with her was the chance to sharpen pencils in her large electric pencil sharpener.

Later, in my twenties, I had spoken to counselors exactly twice, both times about failed relationships. The first time helped, and I never went back, because I'd gotten what I needed. The second time, a few years later, felt pointless, and I didn't bother going again.

Then there was the grief counselor I'd seen on and off after my mom's death, clutching a box of Kleenex in her quiet office. That had been valuable, but, I reasoned, she hadn't been trying to fix anything that was broken in my brain. She'd been a sympathetic ear, a validator of my feelings.

Still. In the two and a bit years since the last crash, I hadn't gotten any better on my own. If anything, I was getting worse. In September 2016, a few months after my night in the Fort Nelson hospital, I pulled over in a rental car on an Oklahoma interstate to weep and hyperventilate in a sudden rainstorm. A few weeks later, at home, while driving to Saturday morning yoga on slick roads after a light dusting of fresh snow, I braked for a red light. I was going slowly, had plenty of time and space to spare, and when I slid, I slid gently to a stop, but my body produced a full-blown fear response even so, just from the feeling that the tires had lost traction. My heart thundered against my ribs; I breathed short and fast through a suddenly tightened chest. When the light turned green, I had to force myself through several long, slow breaths before I could put the car in gear again.

That winter, my friend Maura took a week's vacation to accompany me on a long road trip, doing almost all of the driving herself. I had to go for work, and I was pretty sure I couldn't manage on my own. The next summer, again, I handed the wheel over to a friend, when we found ourselves driving through torrential rain in coastal Alaska. Seeing the water pool in the twin tire grooves worn into the highway brought me to the verge of panic; I just couldn't handle it myself.

In March 2018, nearly two years after the last accident, I bumped into a friend at the grocery store. I was on foot, so she offered me a lift home. It was an icy afternoon, and she drove fast and let the car skid around corners, not quite losing control, delighting her teenager in the back seat. I tried to hold it together. I knew she was just having fun, that we weren't in any real danger. But my chest locked up again, squeezing the air out of my lungs, and I felt dizzy and terrified despite myself. I grabbed for my door handle and held it tight. I tried to keep it in, but suddenly I was crying in the passenger seat, gasping and shaking.

That was the limit for me. Up to then, I had been OK when I wasn't the one in the driver's seat. I'd still trusted my friends to drive safely; it was my own confidence that had been destroyed. But now I was losing it as a passenger? It was too much. I started researching the best therapy options for laying traumatic memories to rest.

I had felt ridiculous for my inability to get past my car accidents. After all, I'd walked away unscathed, or with relatively minor injuries, from each one. But it turns out that car accidents are a common source of trauma, and in fact one in ten people who describe their crashes as "traumatic" will go on to develop full-blown PTSD. I wasn't being silly. My problem was real, and I was far from alone.

After getting advice from a couple of friends who work in mental health, I made an appointment with a Whitehorse-based therapist who specialized in EMDR. The method sounded strange to me, but I liked the physicality of it, the idea that my body would be involved in the cure, my eyes doing the work, rather than relying entirely on my mind: a bodily cure for something I experienced as a bodily problem as much as a *feelings* problem. My fear responses to driving were too powerful, too full-blown and outside my control; I couldn't imagine how I could ever talk myself out of feeling those feelings. I needed to make a more concrete change.

Svenja, my therapist, had started with a standard intake session. She'd asked me about my life, my history, my support network. She needed to get a sense of what kind of shape I was in overall and what kind of resilience I might have while going through the EMDR process.

The second session was devoted to increasing my resilience and emotional resources. Svenja introduced me to the basics of

EMDR and how it came about, and then she offered me a choice. Originally, EMDR had involved the clinician wagging their finger back and forth in front of the patient's eyes, like a metronome, to induce the side-to-side eye movement. But nowadays many therapists use either a set of headphones that emit rhythmic beeps into the patient's ears or a set of handheld pods, sort of like earbuds, that buzz and vibrate in time. I chose the pods, and Svenja let me try them out at different speeds and intensities until I found the settings that seemed comfortable or right to me. Then we worked through a series of exercises, using the pods to drive my eye movement. The idea, Svenja said, was to create a "deeper neurological connection" between me and my emotional resources.

First we established a sort of "happy place" for me: a place, real or imagined, where I felt safe and calm, where, in Svenja's words, "nothing bad has ever happened to you or can happen to you." I closed my eyes, held the pods as they buzzed away in turn, and felt my eyes shifting back and forth behind my eyelids to match their cadence. I chose Fairy Meadows, a place in Nahanni National Park, a remote corner of the Northwest Territories, where I'd once spent a week camping—a place that, in my view, comes close to earthly perfection. I spent a few minutes imagining I was back there: the clear creek running through the green meadow, the marmots whistling to one another, the ragged clouds moving over the face of the mountains, the rocks crashing down off Mount Harrison Smith. . . . It was vivid. I found that with the pods pulsing in their rhythm, and my eyes closed and switching back and forth, I could sink easily and quickly into a potent dream-type state.

Svenja asked me how my body felt, if I felt any warmth or tingling. I told her I could feel my shoulders relaxing. She asked me about my emotions. I felt calm. Then we repeated the same thing again, with her walking me through my feelings and responses out loud.

Three more times we went through the process she called "resourcing." After the "happy place," I thought about a "nurturing figure" in my life while the pods buzzed away. Svenja and I had agreed that using my mother, given my ongoing grief, might be unhelpfully complicating—the resources were supposed to provide unconditionally positive thoughts—so instead I thought about my grandmother, my dad's mom, who'd died when I was eighteen. I pictured her at the kitchen window of her suburban bungalow, leaning toward the window screen with her cigarette; the wrinkles around her mouth and eyes; the clear plastic in her glasses; the colors on the chesterfield in her living room; the smell of Vicks VapoRub and feel of her bony frame when we hugged. The pods buzzed in my palms. My eyes rolled back and forth behind my lids. I felt loved. I felt safe.

After the nurturing figure came a protector. Lastly, I chose a source of wisdom. The point was to give me allies when the time came to fight off intrusive, negative, and traumatic memories and feelings.

Finally, in the third session, we started with the EMDR proper. This time, the pods would be used to open up my bad memories, not just my good ones. After I settled myself on the small couch in Svenja's office, we talked about where to start. I had several incidents to choose from, after all. Facebook's "Memories" feature had just informed me earlier that morning that it was the fourth anniversary of the Jeep crash, so I'd been brooding on that one and how close I had felt to dying. But as we talked things through, it seemed clear to me that the last crash, the rollover in my mom's Subaru, was in fact the worst. Whatever damage had been done before that, the final crash deepened it. It was the one loaded with the most complex and negative feelings, and it was the one that immediately predated all the pulling over and crying. So we decided to start there.

I held the little pods in my hands again, and we fiddled with the settings for a few minutes to try to get the rhythm and intensity just right for me—not too fast, not too slow, not so powerful as to be totally intrusive, but not so faint that I could ignore the pulses. Somehow, I knew when they felt perfect.

Then we began. I closed my eyes, and Svenja had me tell the story of the accident, from beginning to end, while the pods pulsed in my hands, my eyes moved side to side behind my eyelids, and she took notes.

I called up my memories from that day as clearly as I could. I remembered my reluctance to leave the Tim Hortons in Fort St. John that afternoon, the bad feeling I just couldn't shake, the decision to push on anyway. The long evening, the darkening skies, the car coming over that big hill, me squinting to make out the deep bed of hail collected on the road below me . . . Then the car fishtailing, me shouting "Not again!" and the almost perfunctory flip in the ditch.

I talked about the arrival of the various Good Samaritans: the seasonal workers on their way to Alaska, the family who had taken me in. I worked through the surreal hospital visit—those Disney cartoon characters staring at me from the walls all night—and right up to the local family fetching me from the hospital the next morning and putting me on the bus home. I tried to stick to the bare facts of the event, not getting into the grim feelings I had in that hospital room: the guilt and shame over the destruction of my mom's car, the horrible loneliness of realizing that my dad, on a cruise in the Mediterranean, was unreachable, the fear that maybe I was simply and unalterably a truly terrible driver, untrustworthy and dangerous, and that I would just keep crashing until I died. I wondered if I should have explained that mess of feelings more, but Svenja said she'd circle back later if there seemed to be more to dig into.

After I told my story, Svenja asked me to check my body for any kind of reactions. Did I feel tension, pain? I had felt myself getting sad and down as I told the story; I could feel my eyes prickling with potential tears, my lips turning down, my shoulders tensing up. (Antonio Damasio was right once more: It was my body, my physical reactions, that told me about my mood.) Svenja had me hold the pods again and concentrate on those feelings, so I did, until I was on the verge of tears thinking about that grim night in the plastic-covered bed in the children's ward, the guilt and shame and doubt, everything seeming so terrible and impossible and outside my control. It was so vivid, I was almost back there again, miserable and sleepless under the judgmental gaze of Pocahontas.

We went through several rounds of concentrating on the bad feelings, and each time, I closed my eyes and gave in to the rhythm of the pods pulsing in my hands. I felt the tension and pain in my body moving around, like a kind of possession; it went from a tightening in my shoulders and neck to tightness in my chest, escalating eventually to chest pain and shortness of breath, to slight nausea, and then settled in my wrists and hands, which wanted to curl up into fists. Svenja had me put the pods under my thighs instead of holding them, to free my hands, and we went again, the pods pulsing against my hamstrings. My hands tightened, seemingly of their own accord, into fists, and the pain spread up the tendons in my forearms. We both noticed the resemblance to someone death-gripping a steering wheel.

Svenja said all of this was normal and expected, that trauma can be held in the body rather than in thoughts or emotions, and now it was being liberated and ricocheting around. It was the kind of thing I would have rolled my eyes at a few years earlier, but now here I was, feeling as though my body belonged to some alien force. It was a very strange thing to experience. (A survivor of the Oklahoma City bombing who underwent EMDR to resolve her trauma once said that the treatment was "the weirdest thing I had ever experienced, with the exception of the bomb." I've never been blown up, but still, I could relate.)

Next, Svenja encouraged me to come up with a positive statement about the accident, something that I could cement in my brain with the help of the pulsing pods—something like "I did the right thing." But my faith in myself was so shaken, nothing along those lines felt true to me.

"Do I have to believe it?" I asked. Unfortunately for me, I did.

After some debate, we settled on "This was a freak situation," which did feel true to me, along with "It wasn't my fault" (maybe?) and "I'm a good driver" (hmm . . .) and "I deserve good things" (that, I did believe). We did a round dwelling on those ideas, and then we brought in my resources: the peace and quiet of my happy place, the warmth and safety of my grandmother.

My face stopped contorting itself into that about-to-cry expression. The heartburn that had set in during the rounds where we focused on all my negative emotions faded from my chest. My shoulders relaxed. As my body stabilized, my sadness lifted. I felt better.

Two weeks later, I was back in Svenja's office. Just like the last time, she had me settle in on her couch, with the pods set to pulse at my preferred speed and intensity, and then I told the story of the 4Runner crash.

I held the pods in my hands and felt my eyes shift back and forth as I told her about the trip and then what I remembered of the accident itself and its aftermath. The pulsing, the eye movements, the dreamlike meditative state—those things were all the same. But unlike last time, I was not distraught while telling the story of the crash itself. I tried to explain how calm I'd been as the 4Runner spun across the highway, sitting in the driver's seat and just so certain that everything would be fine. I found myself almost laughing at the recollection, how ludicrously wrong I'd been as I sat in my spinning vehicle.

The rollover itself was pure confusion, and I was in shock during the immediate aftermath, staggering in the ditch with my ice-covered glasses and the lump swelling on my head. The negative feelings didn't come until later, after the ambulance ride. I didn't even get upset until my friend Ryan walked through the curtains into my nook in the emergency room to pick me up. Sitting in my hospital gown, I looked up at him, said, "Ryan! My truck!" and started to cry.

But unlike the Subaru crash, which had left me feeling grim and unmoored for days or even weeks afterward, the 4Runner crash hardly touched me emotionally. I remember feeling some fear and worry about my neck when I learned that it was strained. That was about it.

In Svenja's office, I felt mostly fine while we went through the story and experienced none of the dramatic pain and upset of the previous session, just some tightness in my chest, particularly when I talked about my brief stay at the hospital, where I'd had my crying jag. We focused on that feeling in my chest for a round of pulsing, and it ebbed away.

My own calm during the crash likely protected my mind from the trauma of the incident, Svenja said. I hadn't been afraid, so the memories didn't cling in the same way. Something clicked for me when she said that. Of course! That idea made perfect, intuitive sense: Our own feelings of fear and pain can be the cause of the later trauma as much as the event itself is. I thought about how trauma can vary from person to person after the same event: the same IED exploding, the same mass shooting. So much, it seems, depends on those body maps that Antonio Damasio has written about, on the messages that are being sent to our brains: *I am in pain. I am afraid. I am not safe.*

Bessel van der Kolk has described the phenomenon of "inescapable shock." When someone is in a dangerous or threatening situation and they have no escape, no means of acting to secure their own protection, their trauma is inflamed or compounded. Their feelings of helplessness seem to exacerbate the aftermath of the event in their minds. Remember Pavlov's dogs, the ones that nearly drowned? "During the flood," van der Kolk writes, "the caged dogs had been physically immobilized—trapped in their cages—while their bodies were programmed to run and escape in the face of life-threatening danger." It was the fact that they were trapped, Pavlov theorized at the time, that was partly to blame for their debilitating fear later on. I thought back on my memories of the Subaru crash, my useless shouting while I slid toward the ditch. I'd been trapped in the 4Runner, too, but I hadn't really known it. I'd felt confident, in control, because I thought I knew what to expect. I'd been wrong, of course, but in the end my blithe confidence had protected me.

With that epiphany, Svenja and I got through the 4Runner crash fairly quickly. Then we moved on to the Jeep, and things got murkier again. As I had with the 4Runner, I was able to joke and laugh some while telling the story. I remembered the funny parts: the man who'd hit me buying me a halibut burger at my favorite Alaska diner; the strange, almost comical discovery of fine glass shards covering my body. But I also remembered just how lonely and weird I felt in the motel that night in Tok, sensing that I had escaped death and not knowing what that meant exactly, desperately wanting to see my parents, feeling an upwelling earnestness about making the most of my newly salvaged life. And there was no escaping the intensity of the memory of my eyes closing before impact. *This is it.*

Svenja homed in on one thing I'd said: After the Jeep crash, I no longer trusted other drivers. I was always vigilant, I'd said, watching their tires and the yellow center line, waiting to see if they were going to cross over and hit me. The whole system of driving cars—of society, really—relies on a kind of blind trust that everyone else will follow the system, too, and in the Jeep crash, that trust was totally violated. But Svenja tried to get me to see my vigilance in a new way. She pointed out that within reason, a degree of attention and preparedness made me a better driver—wasn't that true? I told her how some of my friends had joked about my "ninja skills" after that crash, congratulating me on my reaction time, on successfully steering my way out of the path of oncoming death. Svenja liked that, and she came up with a set of affirmations for us to go through while I did another session with the pods pulsing in my hands. "I saved myself," she had me say, as my eyes shifted back and forth behind my lids. "I have ninja skills."

It felt silly, but unlike the affirmations about the Subaru crash, it also felt true. She emphasized again that the breach of trust I had experienced would make me a better driver, and I could feel something palpable shift inside me, like a sprocket turning and locking into place.

The day after that session with Svenja, I got in my car and drove north on the Klondike Highway. I was volunteering on the Yukon River Quest, an epic 445-mile canoe and kayak race, and I would be following along by road for the next several days.

Almost immediately, I could tell that something had changed. My dread and anxiety no longer climbed with the needle on the speedometer. I was able to drive at a normal, if slightly conservative, highway speed without repeatedly imagining my own death.

I no longer pictured the car tumbling off the road and into the ditch as I went around every curve. True, the road was dry and the weather was good, but still! I felt better, and calmer, than I had since before the Subaru crash. I looked at the scenery, I listened to a podcast, I relaxed. I had fun.

As the days went by, things got better and better. At first, I was nervous around curves, expecting that sickening surge of fear, but it never came. (Bessel van der Kolk writes that people who have experienced trauma sometimes "develop a fear of fear itself," anticipating that their usual pattern of traumatized behavior will be reasserted. I felt that anticipation keenly, but the pattern I was dreading never reasserted itself.) I was anxious about passing other vehicles, too, having to accelerate to get around big RVs in the short stretches between curves and rises on the two-lane road. But my confidence grew each time I did it. I even drove through some brief rainstorms, and one pelting of hail, and remained more or less calm. I clicked off my cruise control and slowed down a bit whenever I passed through a patch of precipitation, but I didn't feel dread or panic. It was like those reflexes, automatic for the past two years, had been surgically snipped out of me.

I had been liberated from the intrusions of my bad memories, and so I was also liberated from the "anticipatory pain" they had brought me. My lingering trauma was resolved.

I would later learn that the more specific the trauma, the easier it is to achieve relief, and the more broadly the harm is woven into a person's life, the harder relief is to come by. (This is true regardless of the exact method a person selects to try to achieve relief from a trauma.) My crashes were distinct events, and my fearful memories were triggered only when I was driving—only when I was driving on wet or icy or snowy roads, even—and so my trauma was confined to a narrow area where Svenja and I could corral it, round up my intrusive memories, and file them away where they belonged. Other traumas can be much harder to untangle, more insidiously woven into our lives, into eating, sleeping, dating, shopping, watching a movie. I was lucky that mine had been so narrowly defined.

For months after I got better, I could hardly believe it. Even now sometimes, I feel that old anticipation: I'll go around a curve on an icy road and brace myself for the terror to rise up inside me and take control. But it never does.

Impossible as it seems, I was cured.

PART THREE

7

the fear cure

The girl was eleven years old. She was in her bedroom with her mother, who was sewing her a dress. When a mouse appeared and ran across the bedroom floor, and then across the girl's bare feet, she wasn't upset at first; she'd never been afraid of mice. But her mother reacted with raw panic, and the girl, witness to her mother's sudden display of fear, was shocked and distressed.

As the girl grew up, she dated her newfound fear of mice to that day in her bedroom. Years passed, and her fear worsened. She turned twenty, then thirty. She paid obsessive care to preventing mice from entering her home, laying out traps and poison. She tried to avoid places where she had seen mice in the past. At night, she kept a tall pair of boots by her bed, the tops covered to prevent any mice from climbing up the sides and getting in, and if she had to get up in the night, she slipped her feet into the boots and stomped around the house to let the mice know she was coming. When her husband planned a business trip, she made arrangements to stay elsewhere—someplace safe, someplace mouse-free.

Her mother's fear had become her own.

In her thirties, she attempted cognitive therapy. As she entered her forties, she gave EMDR a try. Still, her fear of mice—known as musophobia—persisted, hanging over her, hemming her in.

Three years after her effort with EMDR, the woman approached Dr. Merel Kindt and her associates, in Amsterdam, to ask if they could help her. They agreed to try.

They brought the woman in for just one treatment session: exposure to a mouse, lasting for two minutes and fifteen seconds, and then a single pill. That was all it took. Within a month of the treatment, the woman could get up in the night and walk barefoot around her darkened house. Within three months, she was able to hold a mouse in her hands and even let it run across her bare feet. She was cured of her phobia. She was free.

This story, documented in a 2017 article in the journal *Learning & Memory*, seems far-fetched, too simplistic even to make compelling science fiction. But it's true. Kindt's team cured a woman's three-decades-old fear of mice with a single pill: a forty-milligram dose of propranolol, a common beta-blocker often used to treat high blood pressure, migraines, and performance anxiety. And the banishment of the woman's musophobia was not an isolated case. Kindt has worked her apparent miracle on people with a fear of spiders, a fear of snakes, and an array of other specific phobias.

After years of denial punctuated by occasional meltdowns, after identifying my fear and trying to force my way through it, after my homespun attempt at exposure therapy, I had, in the end, settled for negotiating a truce with my fear of heights, a compromise that I could live with. But when I heard about Merel Kindt's single pill, I wondered, *Could I be truly cured?*

Now and then, when I was a little kid, I used to pass by stretches of sidewalk where fresh concrete had just been laid down. There would be a sheet of plastic stretched over the smooth, wet surface of each new length, and I would always be tempted to lift a corner and carve my initials with the end of a twig. I'd seen the evidence in other sidewalks over the years that this was something kids did, if they were bold enough to make their move before the concrete hardened.

Just before we left Saskatoon, before the move to Ottawa, I saw a fresh patch of sidewalk not far from our house. I remember being especially tempted then, wanting to leave my mark behind as I said goodbye to my home. I don't remember if I actually went through with it or not, but I sort of doubt it. I was, after all, a cautious and inhibited child. I was terrified of being caught and getting into trouble.

The scientific consensus used to hold that our fear memories were like that sidewalk pavement: Once formed, once fear memories had transitioned from short-term to long-term storage, they were fixed, set hard. They could degrade over time, but their basic nature was stable. It turns out, though, that under certain conditions, they can be reopened, becoming labile, or malleable, again. That discovery would eventually form the basis of Merel Kindt's cure.

In the late 1990s, Karim Nader was a neuroscience post-doc working at New York University, under the fear-focused researcher Joseph LeDoux (of the Amygdaloids fame). Nader already knew that our newly formed memories go through an initial period of malleability before what's known as consolidation, the transition to stable, long-term memory storage. He also knew that a large body of research showed that there was a window during which the consolidation process could be interrupted; for instance, that injection of a drug, or the application of electroshock therapy, soon after a round of fear conditioning could disrupt the conditioning process, while the same treatment even just hours or days later had no effect. "One of the most commonly used drug manipulations," he wrote later in an article in *Nature*, "involves the administration of drugs that block the translation of RNA into protein." Disruption of protein synthesis, it seemed, offered a way to sabotage the memory consolidation process. That meant there was a way to stop our fear memories from being internalized.

Nader had also seen research that suggested that those same interventions—drug injections or electroshock therapy—could, if applied when a memory was being retrieved from long-term storage, create a limited form of amnesia; the original piece of learning being accessed would be erased. He theorized that just as memories were consolidated via protein synthesis, memory retrieval might require a similar reconsolidation process, also involving protein synthesis, in order for the retrieved memory to remain intact. There might, he thought, be an opportunity for revision of our fear memories. There might be times when a hard patch of sidewalk became wet again.

He decided to test his theory on rats. He started out with some classical fear conditioning: The rats received an auditory stimulus—a tone—paired with a brief shock to the foot. The next day, each rat was exposed to a single tone and then immediately given an injection to their amygdala. One group received anisomycin, a drug known to block protein synthesis, and the other received artificial cerebrospinal fluid, or ACSF, a neutral, inactive substance.

During that initial presentation of the tone without the shock, both groups exhibited the same freezing behavior, a fear response in anticipation of a shock. But when Nader and his colleagues tested the rats again twenty-four hours later, the freezing response in the anisomycin group had decreased significantly. It was as though the conditioning had been at least partially unraveled.

It didn't work, though, unless the memory of the conditioning had been actively retrieved: A control group of fear-conditioned rats that received the anisomycin without first hearing the tone was unaffected by the injection, and their conditioning remained intact. The results suggested, as Nader had suspected, that protein synthesis was required not just for initial memory consolidation but for reconsolidation after memory activation, too.

In follow-up experiments, Nader and his team found that delaying the injection by six hours after the stimulus nullified its effects; there was a limited window in which to alter the retrieved fear memory. Then, instead of waiting just twenty-four hours after the fear conditioning before administering the tone and the injections, they tried waiting fourteen days. The effect of the anisomycin remained. Here was evidence that, at least in rats, fear memories could not just be overcome by cognitive or behavioral training; they could be fundamentally altered.

Merel Kindt is trained as a clinical psychologist, not a neuroscientist, and she immediately saw the potential applications of

Nader's research. "When I read this paper," she told me, when we first spoke by phone, "I thought, this is really fantastic news for clinical psychology and for psychotherapy." If the process could be adapted for human use, the implications—for people who suffered from fearful memories in all sorts of different ways—could be enormous.

One obstacle: Anisomycin is toxic and can't be used on human research subjects. So Kindt decided to try out Nader's discovery using propranolol; the beta-blocker seemed to have similar properties and was generally safe to use on humans. Propranolol, she knew, had been used by other researchers to alter memories during the initial consolidation process. It was the clear choice.

Kindt and her colleagues started out with a carefully controlled study. "We tested the hypotheses that the fear response can be weakened by disrupting the reconsolidation process," they wrote in a short article in *Nature Neuroscience*, in 2009, "and that disrupting the reconsolidation of the fear memory will prevent the return of fear."

The team used classical fear conditioning techniques to create heightened fear in the study participants where previously there had been none. In this case, they used a loud noise, measuring its effect on the startle reflex by monitoring the muscles in their subjects' right eyes. The next day, they administered a dose of propranolol to one randomly selected group, then they reactivated their subjects' memories of the events of the day before, aiming to open up the fear memories, as Nader and his team had shown to be possible. Another group received a placebo before reactivation, and the third group received just the propranolol without any memory reactivation at all. (Later, Kindt would change the protocol, administering the dose after reactivation.)

The results were encouraging. "In contrast with the pill placebo condition," they wrote, "the administration of propranolol significantly decreased the differential startle response forty-eight hours later."

"Propranolol strongly reduced the expression of fear memory," they went on. "The conditioned fear response was not only reduced but even eliminated." The placebo group did not see anything like the same improvement. The group that received propranolol without the memory reactivation component similarly showed "normal fear responses."

Kindt and her colleagues emphasized in their report that their protocol left intact the memory of the fear conditioning and of the subsequent acquired fear. But, they wrote, "this knowledge no longer produced emotional effects." Kindt's work has sometimes been compared to the movie *Eternal Sunshine of the Spotless Mind*, about a man who receives a treatment to erase the memories of his ex-girlfriend. But the procedure is not about erasing memories. Instead, it's as though it unmoors them so they can no longer trigger our fear responses in the present—a ship untied from a dock, an engine unhitched from a train. In theory, the result is the same as what EMDR did for me, unhitching my memories of past car accidents from my reaction to driving in the present day, freeing me from the fear without erasing it.

Freedom in a single pill is a powerful idea, but it wasn't proven out yet. Next up was testing the method on more powerful, enduring fears rather than on a fear response created in a laboratory just the day before. Kindt and her colleague Marieke Soeter designed another experiment and published the results in 2015.

This time around, their research subjects were forty-five individuals with arachnophobia (as determined by their results on a standardized psychological questionnaire). Once again, the subjects were divided into three groups: One would receive propranolol with memory reactivation; one would receive a placebo with memory reactivation; and one would receive propranolol without memory reactivation.

There was no need for a phase-one fear conditioning this time; these participants were already afraid of spiders. But the team did lay some groundwork. Before receiving their treatment, each research subject was asked to enter a room at the far end of which was a jar on a table containing a baby tarantula. Subjects were asked to approach the jar and to complete, within three minutes, as much as they could of an eight-step standardized behavioral assessment test. They were allowed to stop the test at any time.

First, they were asked to sit in front of the closed jar, just eight inches away. Then they were asked to place the palm of their hand on the side of the closed jar for ten seconds. (If you have a fear of spiders, at this point I expect you're already horrified.) Next, they were asked to open the jar, and then, if they could, to hold the open jar for ten seconds. The steps continued to escalate until number eight, the final task: allowing the spider to walk on their bare hands. If any participant reached this point, they were removed from the research pool. (And fair enough! I don't consider myself particularly squeamish around spiders, but I would struggle with that one.)

Four days later, it was time for treatment. The two groups that were slated for memory reactivation were told that, regardless of how far they'd gotten with the initial test, today they would need to touch a spider in order to complete the process. They were instructed to stand two feet away from a tarantula in an open cage. For two minutes, they remained there—presumably, with hearts hammering, pupils dilated, and all the rest—being asked a set of questions about their fear and anxiety levels and what it was that they feared most about touching the spider. Throughout that time, they lived under the looming belief that they would very shortly be asked to touch the spider. This was Kindt's effort to trigger and reactivate their fear memories. It was a delicate task, bringing them just to the point where the memories became malleable again and not any further.

When their two minutes were up, the subjects were led back outside without having ultimately been asked to touch the spider. Then they were given their pills, either forty milligrams of propranolol or a placebo. Four days later, the participants took the spider phobia questionnaire and completed as much of the eight-step behavioral assessment test as they could all over again.

The changes were stark. Faced with the eight-step test, every single member of the group that had received both the memory reactivation procedure and the propranolol pill was able to progress further in the test than they had before their treatment; many reached step eight and touched the tarantula with their hands. Meanwhile, the groups that had received memory reactivation and a placebo pill, or the propranolol pill alone, remained barely able to touch the jar. Later, at three-month and one-year follow-ups, the group that had received the full treatment remained stable; they hadn't regressed. They could still touch the jar.

When Kindt and I first spoke on the phone, just a little over three years after the publication of the arachnophobia study, I wasn't sure of the current status of her research. I wasn't sure if she was taking new subjects or if I would be a suitable candidate. But then, near the end of our chat, she rattled off a list of phobias she'd tackled so far: spiders, snakes, silverfish, dogs. Confined spaces. Heights.

"That's what I have," I said.

"You want to be treated?" she asked me, laughing a little. I told her, also half kidding, that I would love to try.

But I was in luck. She had opened the Kindt Clinics a few months earlier. She was now treating people not just in the course of her ongoing research but as regular patients. For one hundred euros an hour, if I met the criteria to be accepted, the cure could be mine.

A few days later, I filled out an online questionnaire, supplying answers about my fear and rating my level of agreement with statements like "I avoid having to face the situation at all costs." I got a doctor to check my blood pressure and sign off on the dosage of propranolol I would have to take; since the drug is a blood pressure suppressant, it isn't necessarily safe for anyone whose blood pressure is already too low.

One of the questions was about my family history. Had anyone in my family died of a heart attack or related causes by the age of sixty? I disclosed my mom's stroke at sixty and her father's death at fifty-four from an aortic aneurysm, and as I typed my answers into the form on my screen, I was startled all over again by the bald facts of how unfair my mom's life had been.

But regardless of that sad history, the clinic accepted me for treatment. I booked a flight to Amsterdam and tried not to get my hopes up. Kindt's results so far have been strong. But the success or failure of the treatment seems to hinge almost entirely on the reactivation component—the part where the sidewalk concrete is made wet again. The challenge of the treatment is in retrieving the fear memory in such a way that it becomes labile.

"Why do we have memory?" Kindt asked me when we spoke, before answering her own question. Memory's essential purpose, she said, is to help us adapt efficiently to our environment, to learn about threats and then retain that information without having to relearn it each time a given threat appeared. That purpose helps explain why fear memories are generally static and enduring: They'll be needed as warnings in the future. They only become labile, alterable, again if there's good reason for them to open up.

For that to happen, Kindt explained, "there should be something new to be learned, otherwise the memory trace is only in a sort of passive state retrieved, but the memory trace doesn't open up." On the other hand, if the threat that the person is exposed to in treatment is too new, the brain will create an entirely new memory, a new piece of learning, instead of revising the old one. She offered an example: If someone were afraid of spiders and you exposed them to a spider for half an hour, or an hour, or several hours, their initial wave of fear might eventually begin to fade, at least to a low simmer. "If you do this, then a new memory is already formed," she explained. After that, the propranolol would affect the newly formed memory of the low-simmering fear rather than the original, more potent memory.

That person would in all likelihood have to start all over again the next time they were exposed to a spider. Kindt's goal was to prevent the appearance of the fear response at all, and to do that, she had to trigger people at just the right pitch. I've never been a huge fan of analogies that compare the human brain's agile, endlessly creative workings to the cold functioning of technology, but this one seems apt: What Kindt needed to do for her patients was to open up a given memory file in edit mode, not in read-only mode, without prompting the computer to draw up a whole new document instead. It was a difficult needle to thread.

"The difficulty in translating the basic science to clinical practice is that we cannot directly see what's going on in the brain," she told me. She has no surefire way to know, while working with a patient, which of the following processes she may be observing at any given time: "This is passive retrieval, nothing is happening; this is probably reconsolidation; and this is already, wow, this is too long, and then we passed the window of opportunity to change the fear memory itself." Instead, she has to go partly by intuition, by asking questions of the patient, by observing their reactions to the process. When she hits the bull's-eye, the drug seems to do its job. But the reactivation is still an uncertain process.

On the plane to Amsterdam, I tried to imagine what my life would be like, feel like, with my fear of heights neutralized. I couldn't really grasp it—and I didn't want to try too hard, since it was possible that the treatment wouldn't work for me. I didn't want to conjure up a vision of a life free of this fear, fall in love with it, and then have it slip away.

I'd been encouraged by my EMDR experience. Change, even dramatic change, in my relationship with fear *was* possible, more possible than I could have believed before I walked into Svenja's office on that sunny summer day. But at the same time, my traumatic memories from the car wrecks had always felt like an intrusion, an invasion of my mind, something that had been stapled onto my life from the outside. As powerful and intractable as my fear reactions had seemed when I experienced them while driving, they had also felt alien to me. I could remember so clearly how much I had enjoyed driving in my pre-crash life. The bad feelings were foreign, parasitic; it was logical that they could be expelled eventually.

My fear of heights, though—that was different. It was part of me, woven into my life since that day I fell down the escalator at Pearson Airport as a toddler, and quite possibly even before that. It wasn't that I wanted to keep it, that I felt attached to it emotionally; I just couldn't fathom how it could possibly be severed from the rest of me.

I pictured myself cured, the panics in high places as mere memories rather than premonitions of future panics to come. I imagined them receded into a fearful past, as untouchable to the new me as my good memories of driving had been while the trauma had its grip on me. Things would be different in small but meaningful ways, I figured. I wouldn't have to ask around about the terrain before going on a hike anymore, checking its suitability and gauging the likelihood of my winding up sobbing, frozen in place, humiliated and incapacitated by fear. I could become a bolder, better version of myself.

In my intake interview, conducted over FaceTime with Kindt's colleague Maartje Kroese a couple of days before I flew to Amsterdam, I'd been asked what my hope was for the outcome of the treatment. I'd said that I doubted I would ever be someone who enjoyed exposure to heights—I didn't imagine myself becoming a regular skydiver, for instance, or an ace rock climber—but I hoped that the treatment would silence the voices in my head that rose up, in perfectly normal situations, to scream *You are going to die!* It was the outsized, irrational reactions that bothered me the most. Some fear, some discomfort, I could live with. If, while hiking on an exposed slope, I could instead think, *A fall here would be a bit uncomfortable, I might get scraped up a bit*, that alone would be worth the trip.

Ahead of our interview, I had sent Kroese some of my writing about my fear of heights. During our forty-five-minute chat, we went over some of the lowlights of my career in fear-panics. We talked about the escalator, and the top of the Duomo in Florence. We talked about my descent from the Usual. We talked about what drove me to resist practicing avoidance—why I continued to expose myself to situations that I knew risked triggering my fear—and we talked about how my body felt when it was triggered.

Kroese and Kindt needed to understand exactly how my fear worked in order to activate my fear response in such a way that it could be altered. It all hinged on that. Kroese had got me to rate my past fear-panics on a scale from 1 to 100. In the treatment, she told me, they would try to get me up to an 80 or 90.

We talked a bit about the methods they might use in a controlled environment. The Netherlands, after all, is not known for its vertiginous terrain; mountain exposure seemed out of the question. We talked a bit about ladders, about rock climbing gyms, about the steep, narrow stairwells of ancient churches. Kroese discouraged me from dwelling too hard on the possibilities. Leave it up to them to figure out the answer, she said. I should try my best not to think about it.

So as I flew from Vancouver to Toronto to Amsterdam's Schiphol Airport, not only did I try not to think about whether the treatment would work, I also tried not to think about what they were planning to do to me.

Somewhere over Ireland, with dawn just lightening the eastern horizon, I realized something. This flight, from Toronto to Amsterdam, through a night shortened by our traversing of time zones, was the first red-eye I'd taken since the terrible night when I rushed from Seattle across the continent to my mother's hospital bed. As I sat cramped in my window seat, flimsy airline blanket pulled up to my neck, the memories of that night in 2015 flooded back, as bad memories do when they're released from storage: How I'd pressed my forehead against the plastic of the window, hiding my tears from my seatmates. How I'd chanted the truth over and over to myself. *Your mom is dying. You will never talk to her again. She's already gone.*

The next day, I met Kindt and Kroese just before 11:00 AM Amsterdam was rainy and grey, the wind blustery. Back at home in Whitehorse, and in my mind and body, it was 2:00 AM, but after my flight, in the quiet studio I'd rented near the clinic, I had managed a solid eight hours of sleep, and I was feeling OK. Before bed, wrecked by jet lag, I'd been anxious and overwrought about the treatment. Would it work? How cruel would it be if I wasn't scared enough for it to function? What if, at the one moment in my life when I wanted the fear to overtake me most fully, it refused to perform?

But so far this morning, I felt calm. All I could do was put my faith in the clinicians and see what happened.

I had agreed to be filmed during my treatment by a Canadian documentary crew. They were making a TV special about the science of fear and by sheer coincidence had planned to be in Amsterdam to film several of Kindt's clients when I would be there. They met me out front of the clinic building and shot several takes of me approaching the doors. I strolled. I stared into the middle distance. The repetition was unexpectedly soothing—a good distraction from what was coming.

Inside, I met Kindt for the first time. With a camera rolling, we sat down for a brief interview. We talked about the particular ways my fear seemed to be triggered; we talked about where my fear lived in my body when it was active. She took my blood pressure and asked me how I felt, fear-wise, on a scale from 0 to 100. I had felt traces of the old panic rising in my chest as I told her about the incident on the Usual. I told her I was already at a 30.

Then she unveiled her plan for me. We were headed to a fire

station. I would climb into the bucket on a ladder truck and rise into the sky. What did I think? I burst out laughing. "I don't think I'll like that at all," I told her.

We all piled into the film crew's minivan for the journey to the outskirts of Amsterdam. At first I joined in the chatter in the vehicle, but soon I quieted down and stared out my window at the flat green fields. I was nervous. I was afraid of what was about to happen, but also, more powerfully, I was afraid that I wouldn't be scared enough. I hadn't had a full-blown panic in more than three years—not since the Usual—and I was afraid I'd gotten too good at suppressing my reaction, at controlling it to some degree, even while I was unable to free myself from it. Kindt had told me to do my best to let go, to let it rise up and overpower me, to drop my defense mechanisms. I hoped I would be able to do so.

I waited outside the fire station while the others went into a courtyard to set up. Just over a fence, ducks paddled in a suburban pond. I tried to think about how I'd felt in Florence, trapped by that inescapable mental image of my body sliding over the terracotta tiles of the Duomo, or when I was halfway up the tall ship's mast, on Lake Ontario, swaying and frozen, fixated on the image of my body shattering on the deck below. I tried to let the memory of that fear flow through me.

Finally it was time. Kindt led me to the ladder truck's bucket—more of a small fenced platform, really—and had me get in and stand in the corner next to a gate in its safety rail that could be opened to increase my sense of vulnerability, if I needed to be terrorized further. A tall, round-faced firefighter was in the bucket with me, at the controls; another sat at the wheel of the truck itself. They both made eye contact with me and smiled, and I wondered how this ranked in terms of weird uses of their time. Probably pretty low, I supposed, all things considered. Kindt stood at my shoulder; the camera operator, packed in with us, too, peered around the firefighter's bulk. *Can this thing hold all four of us?* I thought. And then we began to rise into the air.

Instinctively I reached out to hold on to the rail; Kindt made me release it. I stuffed my hands in my pockets; she made me give that up, too. "Safety behaviors" were forbidden. Now only my feet connected me to the platform, which jerked and shivered in the wind. "Oh my god," I said. "Holy shit." We rose higher. I forced myself to look down—at the courtyard shrinking away from us—instead of steadying myself by looking out at the flat landscape stretching to the horizon. The firefighter turned us around in a slow circle, and the courtyard spun below me. I felt sick, dizzy, terrified. I told Kindt I was at a 60.

We rose higher still, until the truck was maxed out, until we stood well above the fire station, higher than anything else visible in any direction. The wind blew my hair into my eyes, shook the platform from side to side. I moaned. I didn't feel like I was going to die, but I was *not* happy. Was it enough? I told Kindt I must be at a 70 now. Time passed—I don't know how much. Not more than a minute or two, likely, though it seemed far longer. I was vaguely aware of the camera guy zooming in on my face as I groaned and my hair whirled around me. Kindt stood close and looked into my eyes, trying to gauge the level of my terror.

Was it getting any easier for me to be up here? she asked eventually. I was getting used to it, a little, I said, when the wind wasn't gusting and the platform was still. That was the signal for us to head down and hope for the best.

Back on the ground, my legs shook. I took my time exiting the bucket and stood on the pavement of the courtyard, feeling ready to collapse. My chest was tight, my breath short and fast. I gulped for air. Kindt brought me a bottle of water and a single pill. The propranolol. I swallowed it, then went to wait, shivering, in the shelter of the van while the camera crew gathered some additional footage. Kroese came with me and told me not to think anymore about what I had just done. We made small talk with the van's heat on full blast.

Sooner than I expected, I felt completely calm. My body felt normal again. My heartbeat was slow and steady, and the shakiness was gone from my legs and my chest. That was the pill, Kroese told me.

At the clinic, I read a book in a quiet room for a couple of hours. Kindt and the film crew left, bound for a farm where another patient would confront their fear of chickens. Kroese took my blood pressure again, then sent me home with a reminder not to think, talk, or write about what had happened until I'd gotten a good night's sleep. I told her I'd try my best.

The next afternoon, I was back at the clinic, back in the film crew's minivan, back at the fire station. I'd gone to bed early but had woken up after four or five hours of sleep; it was, after all, the middle of the day back in Whitehorse. I hoped I had offered my body enough REM cycles to get the job of memory reconsolidation done.

I felt nervous, but it was nervous excitement, not nervous dread. I searched inside myself for any creeping fear of what was coming—and I didn't find any.

All morning, I had tried to look for any sign of change. Was I brasher, less nervous, on the steep stairs that led up to my rented apartment? I thought I might be. But then, had the stairs really scared me before anyway? It was hard to say.

I stepped onto the platform with Kindt and the camera guy again, and I was not afraid. A different firefighter was at the controls. Butterflies swirled in my stomach, but, again, they felt like the fluttering, anticipatory kind, the stirrings that precede a date with a crush, or before you stride out onstage to accept your hard-earned diploma.

The crowded bucket began to rise into the sky, and I was not afraid. I looked down at the courtyard receding below us and felt none of the nauseating, dizzying terror of the day before. I was fine! I grinned, then started to laugh out loud.

We rose higher. The wind shook the bucket from side to side, and I laughed. I stared down at the ground, then out at the horizon, then down at the ground again, and found that my only worry was that my glasses might fall off my face and plummet to the pavement below. I rummaged around inside of myself. Where was my fear? It should have swelled up and taken over my body by now. Was it hiding in there somewhere, waiting to pounce if I let my guard down? But I couldn't find it.

Kindt asked me a few questions; the camera guy chimed in, too, holding his lens close to my face. I don't remember what they asked me or how I answered—I remember only my sheer giddiness at the sudden, unexpected absence of my fear.

As we neared the ground again, I joked that I was smiling too much. The camera would catch the yellowing of my teeth, I said, feeling suddenly vain and self-conscious. Yesterday I'd been too terrified to care how I looked, I realized. Without the fear, there was so much more room in my head. Looking at it that way, I almost had to welcome the spike of insecurity.

Back on the ground, I still wanted to laugh, still couldn't stop smiling. The film director asked if Kindt and I would be comfortable going back up again, with the camera guy on the ground this time, so he could get some footage of us rising up into the air from that perspective. "Sure!" I said, brash, savoring my lack of terror. I could go back up again. I could do anything.

Kindt and I stepped back into the bucket and rose into the air. Still my fear stayed away. We went even higher this time—I guess because we had fewer pounds of person in the bucket—and the wind seemed even stronger. I took a selfie with Kindt, quickly, worried that my phone would be blown right out of my hand. Now and then, I looked down at the ground, to prove to myself that I still could, but mostly, this time, I tried to enjoy the view. Soon, though, I was distracted. The firefighter seemed nervous about the wind and the height and the time the camera crew was taking to get their shots, and I felt the first tinglings of fear creeping back in. I started to worry—not so much about our safety but about my cure. Would it hold under this strain? My elation drained away, and I shivered in the cold March wind. "I think I'm done with this," I said to Kindt.

By the time we reached the ground, I was shaken up. Not panicked, not like the day before, but uncomfortable, my mood sinking.

The camera guy and the director met us as we stepped off the platform. How was that? they asked. I answered something like "not great." They tried to follow up, camera rolling, and I told them I needed a minute to collect myself. They kept filming, standing in front of me, still tossing questions my way. "I just need a minute and then I'll be happy again," I said, and Roberto, the director, said that they didn't need me to be happy for the camera—they just wanted to capture the reality of what was happening with my treatment.

You know that panicked feeling when your throat tightens up with anger or upset and tears rise to your eyes, and you try to hold yourself together, but the harder you try to control your emotions, the more they try to explode out of you? I was angry. I was disoriented and shaking from the cold. I just wanted a moment to take a deep breath and pull myself together! Why wouldn't they listen? I was being as clear as I could manage. I felt cornered, disrespected, antagonized. Most of all, I was afraid that my explosion of negative feelings might compromise my cure. They wanted to capture my reality? My reality had been elation, freedom, joy. What I was feeling now was a result of my efforts to accommodate their needs, their demands.

Finally I said, "I'm freezing, I can't even think, I can't do this right now," and walked away from the camera.

I was mostly silent on the drive back to the clinic. I tried to call up my memories of the sheer joy of the first successful bucket ride, but my angst and anger seemed to have overpowered them. What should have been a triumph, a pure victory over decades of fear and shame and tears, felt ruined. When I made it back to my room, I collapsed on the bed and sobbed, letting out all the sadness and anger that I'd held back until I could be alone.

When I'd calmed down some, I thought back to Svenja and my resources. I tried to conjure Fairy Meadows, its raw rock faces and ragged clouds, the green meadow and the cold, tumbling stream. I pulled up a vision of my grandmother, her wrinkled eyes and cheeks, her bony Vicks VapoRub hugs. My mood eased, a little.

I emailed Kindt and Kroese for reassurance. Could getting as upset as I had, immediately after coming out of the bucket, compromise my cure? I had learned enough, in my efforts to understand my fear, to respect how powerfully emotion and memory rule us.

But it didn't work that way, they said. The change had been made. And besides, Kindt reminded me, *some* fear was natural and good. Some reactions were reasonable, not irrational at all. She'd been afraid herself during that second bucket ride, she told me. It seemed clear to both of us that the firefighter had been nervous, too. It had been so windy; we'd been up so high. Anyone could have felt rationally unhappy in that situation. And here was something else for me to remember: I had lost the joy of the first successful try, sure, but I hadn't panicked the way I had the day before. I hadn't even come close to experiencing that full-body, uncontrollable terror.

I realized I would have to relearn my own reactions. For so long, I had worked at suppressing and ignoring my fear responses. I had taught myself that they were irrational, not to be trusted. Now, if I was truly cured, I would have to learn to trust them again.

8

fearless

Alex Honnold clung two thousand feet above the Yosemite valley floor, a human speck on a sea of granite.

It was September 6, 2008, and Honnold was attempting to free-solo the Regular Northwest Face route up Half Dome. In 1957, when the route was first established, pioneering climber Royal Robbins and his team spent five days laboring with ropes, bolts, and pitons to reach the top. Fifty-one years later, Honnold was looking to do the same route in a matter of just a few hours, alone, with no ropes or any safety gear whatsoever. It would become the first in a series of iconic, ropeless, big-wall climbs—culminating with the ascent of El Capitan, or El Cap, documented in the Oscar-winning film *Free Solo*—that would make Honnold the most famous rock climber and one of the most famous outdoor athletes in history.

Honnold had started out that morning on Half Dome wearing nothing but shorts, a long-sleeve T-shirt, his rock shoes, and a chalk bag strapped around his waist. He had a few Clif Bars in one pocket, a small flask of water in the other.

He had climbed up and up and up. Drank, ate, dipped his hands into the chalk bag. Pulled off his T-shirt somewhere along the way. As he put hundreds of feet of rock behind him, he felt fear, or something along the fear-anxiety spectrum, a handful of times. "Holy shit!" he thought at one point, about a thousand feet up, as he realized he'd worked his way off the established route by accident. "This is hardcore." But, as he wrote later in his memoir, *Alone on the Wall*, "It wasn't true panic that I felt—just an uncomfortable anxiety."

He'd focused his mind, rallied, and found his way out of that jam. Then, hundreds of feet higher, just a few dozen feet from the top, he'd come to the last tricky bit of slab climbing. This was where he stalled out.

"I had a moment of doubt," he wrote years later. "Or maybe panic. It was hard to tell which."

As he paused there, clinging to the rock, he alternated his hands on a single "pathetic ripple" of a handhold—holding with one while he rested the other. His feet weren't on proper holds at all; he was using a technique called smearing, relying on the oppositional forces of his rubber-soled shoes angled and pressed hard against the granite to maintain traction. He could hear tourists at the top, laughing and chatting just above him. His calves burned from the pressure of holding himself in place. Minutes ticked away. He had to move soon.

Finally he stood tall, reached out, and grabbed the next hold, the one that had seemed tenuous. His feet held. His hands held. He had made it through, and moments later he pulled himself over the top of the wall and found himself in a crowd of a hundred tourists—shirtless, panting, and totally anonymous. No one at the summit even realized what he'd done.

Later, in his journal, Honnold noted that he had completed the climb in two hours and fifty minutes but that he was dissatisfied with his performance. "Sketchy on slab," he wrote, with a sad-face emoji. "Do better?"

I've followed Alex Honnold's career for years, and while I find him fascinating, it's not often that I find him relatable. We are, to put it mildly, very different people. But when I was rereading his description of that brief stall on Half Dome and its impact on how he viewed the whole climb, I thought back to my own efforts to conquer my fear of heights during my summer of DIY exposure therapy: Sending the route didn't count toward a cure; there was no victory, if I couldn't train my brain to stay calm.

Honnold likes to insist that he's really a pretty normal person, fear-wise. "I feel fear just like the next guy," he writes at one point in *Alone on the Wall*. "If there was an alligator nearby that was about to eat me, I'd feel pretty uncomfortable." "Pretty uncomfortable" is a milder word choice than most people are likely to use for that scenario, though.

"I get asked all the time about risk," he writes. "The usual questions are 'Do you feel fear? Are you ever afraid? What's the closest you've ever come to death?' I get really tired of answering those questions over and over again."

Fair enough. But the questions are understandable. From what we, the public, get to see of Honnold's professional life, his relationship to fear isn't anything like most people's. It certainly isn't like mine. His ability to execute precision climbing moves, for hours on end, when any mistake would result in his sudden and certain death, is barely possible even to comprehend. He seems immune to the shaking and sweating, the hammering heart and the constricted airway, that would find most of us trapped in such a situation.

Still, there are other arenas where his responses seem more normal. Like me, and like Dr. Mujica-Parodi and all her sweaty research subjects, Honnold has tried skydiving. Thinking that he might eventually advance to BASE jumping—parachuting from a cliff or tall structure—Honnold jumped out of a plane for the first time in 2010, one of a handful of dives he completed before giving up the sport. I felt a satisfying sense of recognition, of validation, when he wrote that he "hated everything about them."

"I felt vaguely motion-sick on the bumpy plane rides up, crammed in with the other jumpers like sardines and breathing exhaust fumes," he writes. "And I found falling out of a plane to be just plain scary."

Alex Honnold! He's just like us!

Or maybe not. A couple of years ago, the writer J. B. MacKinnon persuaded Honnold to climb into an fMRI machine so that neuroscientist Jane Joseph could study his brain. A story had been circulating about a neurobiologist, waiting in line for an autograph at one of Honnold's public events, who'd leaned over to his neighbor in line and muttered, "That kid's amygdala isn't firing." Joseph planned to see if that armchair diagnosis was correct.

MacKinnon's story tells what happened next:

> An initial anatomical scan of Honnold's brain appears on MRI technician James Purl's computer. "Can you go down to his amygdala? We have to know," says Joseph. . . .
>
> Purl scrolls down, down, through the Rorschach topography of Honnold's brain, until, with the suddenness of a photo bomb, a pair of almond-shaped nodes materialize out of the morass. "He has one!" says Joseph, and Purl laughs. Whatever else explains how Honnold can climb ropeless into the Death Zone, it isn't because there's an empty space where his amygdala should be. At a glance, Joseph says, the apparatus seems perfectly healthy.

But she wasn't done yet. Inside the machine, Honnold had been instructed to look at a series of images designed to provoke fear, distress, disgust, or alarm in the viewer. There were images of bloody corpses, images of feces, images of . . . mountain climbing.

On the glowing screen showing Honnold's brain activity, the amygdala did not light up to show that it was triggering a fear response. "Maybe his amygdala is not firing—he's having no internal reactions to these stimuli," Joseph said. "But it could be the case that he has such a well-honed regulatory system that he can say, 'OK, I'm feeling all this stuff, my amygdala is going off,' but his frontal cortex is just so powerful that it can calm him down."

Joseph had scanned a control subject, too: another rock climber about Honnold's age, someone who would also be classed as a "high-sensation seeker." Like Honnold, he had seemed, to his observers' eyes and by his own account, largely unaffected by the images as he looked at them. But his brain scan had told a different story: His amygdala was active, even if he wasn't consciously perturbed, or admitting that he was.

What to make of it? Honnold's amygdala is technically functional. He seems, though, to have a higher threshold than most people do for its activation, likely through some combination of nature and nurture—an inherent variation in his responses further honed to a fine, fearless edge by his years of discipline and training and exposure to high-risk situations. It's clear that his relationship to fear, his response to potential threats, looks different from the average human's—and different even from that displayed by his fellow seekers of danger.

"Danger scares me," Honnold wrote in his memoir, after relaying the story of the autograph-seeking scientist who figured his amygdala must not work right. "But as I've told countless folks who ask, if I have a certain gift, it's the ability to keep myself together in places that allow no room for error. I somehow know, in such a fix—like the moves above Thank God Ledge on Half Dome where I stalled out in 2008—how to breathe deeply, calm myself down." It was that same kind of calm, I realized, that I had been chasing during my exposure therapy efforts.

It wasn't that I had ever hoped or expected to reach a place where rock climbing felt as ordinary, as neutral, to me as sitting on the couch or walking along a flat, smooth path. I didn't want to be oblivious to my surroundings and the risks they might present. But I wanted to be able to rise above them, to breathe through them, to prevent the panic from swelling up through my chest and overtaking me.

I suppose that ability, that resistance, is a kind of fearlessness, and Alex Honnold's extreme version of it has dazzled viewers around the world. But there's another kind of fearlessness, too, one that's created when a person's amygdala—unlike Honnold's—is truly, totally nonoperational. Somewhere in the American Midwest, hidden away from movie cameras and magazine photographers, there is a woman living something remarkably close to a life free of fear. She is well known to neuroscientists and fear researchers as Patient S. M.

One day in the mid-1960s, a baby girl was born who didn't scream or cry. She barely even whimpered, but her silence wasn't due to a lack of feeling: The baby had a strange thickening in the tissue around her vocal cords, and with that obstruction it was difficult for her to produce any sound.

She had lesions on her skin, too, and those—combined with the trouble with her vocal cords—eventually led to her being diagnosed with an incredibly rare genetic condition, Urbach-Wiethe disease.

Urbach-Wiethe attacks people on two fronts: in their skin and throat tissue, and in their brains. Beyond the characteristic thickening of the tissue around the vocal cords, which gives Urbach-Wiethe patients a distinctively hoarse, scratchy speaking voice, the disease also causes skin lesions, generally on the limbs, which lead eventually to widespread scarring. On the neurological front, Urbach-Wiethe causes calcification in the brain structures it infiltrates, and sometimes the damage is extensive enough that it effectively knocks those structures offline. Only a few hundred people have been identified as having the condition.

A number of diseases will attack our brains with what one neuroscientist calls "creepily specific lesions," damaging only certain brain structures for reasons we don't entirely understand. Herpes simplex encephalitis is one example; rabies, most famously, is another, targeting the hypothalamus. Urbach-Wiethe disease, for its part, appears to have a particular predilection for the amygdala.

So the baby girl grew up with a strange-sounding voice and waxy-looking, scarred skin. She was treated about as well as you would expect by other kids and, as you might also expect, wound up feeling unattractive and alienated. Then, when she was around age ten, the disease began to creep into her brain.

One day, when she was still a small child, before the disease did its work on her neurons, she went for a hike in the woods with her father. Walking through some loose brush, she fell several feet into a large pit that had been partially hidden by fallen branches and leaves. The pit was home to a nest of baby snakes, and as they slithered, startled, around and over her, sliding across her legs, she screamed and screamed to her father for help. Her remembered terror was still vivid even decades later. But that was likely one of the last times she felt that kind of fear.

Years passed. She had her DNA sequenced by a doctor in Switzerland; later, she had regular laser surgeries to pare back the excess tissue around her vocal cords and prevent a dangerous obstruction of her airway. She turned eighteen. She had her first sexual relationship and got pregnant, and the man left her when he found out about the baby.

The young woman entered her twenties and wound up in an abusive relationship with a new man with whom she had two more children. The father of her second and third children left her during her last pregnancy, and a couple of years later, now a single mother of three, she married. The marriage lasted less than a year before collapsing when the woman confronted her husband about his infidelity. The confrontation ended with the man strangling her until she blacked out. When she woke up, he was gone.

The baby who didn't cry, the girl with the funny voice, is now a woman in her fifties. Her life story is studded with grim facts like these. She has been assaulted and had a gun held to her head. She has experienced more harrowing and violent events than many of us endure in a lifetime. But if you met her and she told you about them, she would describe them to you calmly, without showing any sign of fear.

The woman, who would become known as Patient S. M., had her first brain scan when she was twenty years old. Our technology has come a long way since that first scan in 1986, but even back then, the damage was striking: two pale, bean-shaped patches in the darker mass of her brain matter, mirroring one another in each hemisphere; voids where her two amygdalae should have been. In the decade since it had begun to do its work, the disease had destroyed the structures virtually completely while leaving practically everything else untouched, one of those "creepily specific" excisions, the kind of thing the scalpel-happy psychosurgeons of the past might have, on some level, envied or admired.

In "A Tale of Survival from the World of Patient S. M.," published in the book *Living Without an Amygdala*, the neuroscientists Justin Feinstein, Ralph Adolphs, and Daniel Tranel describe S. M.'s lesions as "the most complete amygdala lesions ever reported" from Urbach-Wiethe disease. The damage was "unlike anything that had been seen before."

That first scan came about after S. M. was referred by another neurologist to the clinic at the University of Iowa. There, Antonio Damasio and his wife, Hanna Damasio, had recently launched a registry of neurological patients with brain lesions, and S. M. joined the registry as patient number forty-six. She made herself available to researchers interested in learning what they could from her, and plenty of them took her up on the offer.

"She was sort of the star patient, based on the number of *Nature* papers that she generated," Adolphs told me when we spoke by phone, laughing at the memory of the piles of work she inspired. (Studies of S. M. have generated dozens of peer-reviewed publications that have, in turn, been cited more than thirteen thousand times.) Adolphs arrived in Iowa in 1993, as a postdoctoral fellow studying under Antonio Damasio, and his work there over four years changed the course of his career.

"It was really just serendipitous," he said. Adolphs had been studying cognitive neuroscience in his earlier work, and when he got to Iowa, there was the Damasios' registry, just waiting to be used. He dug in. "I guess it was pretty much random that I happened to study some that had interesting dissociations with respect to emotion, like Patient S. M. . . . And then that sort of launched my career in social and affective neuroscience."

Studies of patients with brain lesions like the ones in the Damasios' registry are valuable because the absence of something can be instructive about its role when it is present. ("Neurology's favorite word is 'deficit,'" Oliver Sacks once wrote.) Think of a car engine. If the car is running fine, and the engine is intact, there is only so much that an outside observer can deduce about how it works. But if you remove a single part and carefully observe the effect its removal has on the operation of the whole (and then repeat this carefully observed removal with various other single parts, systematically, one at a time), you might eventually start to understand things better. In S. M.'s case, her lack of a functioning amygdala has taught us a lot about the role that brain structure plays in the overall system—and, more broadly, about what it even means for us to feel fear.

Fear, or the lack of it, quickly became the focus of the studies done on S. M. People tested her ability to recognize fear on the faces of others (she can't see it, although she can perceive their sadness or anger). They also studied her inability to pick up on bodily cues suggesting fear or danger. They tested her physiological fear responses—her reaction to loud noises, for instance, like the one that terrorized Little Albert a century ago. They scrutinized her decision-making, and her relationship to risk-taking and rewards. They did their best to figure out, by its absence in S. M., exactly what role the amygdala plays in the car's engine.

Then, in 2003, nearly two decades after they'd begun studying her, the researchers at Iowa decided to broaden their approach. They launched a multiyear study with one central aim: to see if they could frighten their apparently fearless research subject.

More specifically, they wanted to understand the amygdala's role in the *experience* of fear. An array of studies had established that the structure is involved in fear-related functions. We know it plays a role in fear conditioning, for one, and in the triggering of physiological fear responses. We know that it gets involved when our breath shortens and our attention to a possible threat narrows and intensifies. We know it has a role to play in all of that. But what about the feeling of fear itself?

They set out to expose S. M. to a series of stimuli that you'd expect would trigger some degree of fear in the average person. They judged her reactions on two fronts: by monitoring her for typical fear behaviors (jumping back in startlement or screaming, for instance) and by asking her to complete self-reports about her feelings during each round of testing. They predicted that they would be unable to induce fear in their subject—but even so, her imperviousness surprised them.

They started with snakes and spiders, creepy-crawlies whose proximity makes many people, even if they aren't phobic at all, at least mildly uncomfortable. Some degree of discomfort around these kinds of creatures is deep-seated, honed by millions of years of evolution. And besides, S. M. had told them the story of the hike, the pit, and the baby snakes slithering everywhere while she screamed. She had told the researchers repeatedly over the years that she "hated" snakes and that she did her best to avoid them. Though she was now apparently fearless, the memory of her childhood terror in those horrible moments in the pit remained intact.

So when Feinstein, Adolphs, and the rest brought her to an exotic pet store, they expected that she would avoid the snake section. She loved other animals, and they figured she would focus her attention on the hamsters and the puppies, and that they might even get a glimpse of some nervous or fearful behaviors if she came close to the snakes by accident.

They were wrong. When S. M. and the researchers entered the store, she headed straight for the snakes, fascinated, peering at them closely through the glass. Noting her interest, a store employee removed a smaller, nonvenomous variety from its enclosure for her to hold. S. M. let it wrap itself around her hands. She stroked its scaled skin; she touched its flickering tongue. "This is so cool," she said. Later, she asked fifteen different times if she could try holding the larger, more dangerous snakes. (The store's employees refused her each time.) It wasn't just a matter of her feeling safe in the confines of the pet store either. Later, a family member told the research team that she'd once tried to touch a snake she encountered in the wild, too.

So much for her fear of snakes.

It was the same with the store's tarantula. She had told the researchers that she avoided spiders, but now she begged to be allowed to hold the furry menace.

Next, the team took S. M. for a visit to the Waverly Hills Sanatorium, in Louisville, Kentucky. Once home to dozens of tuberculosis patients, the place now bills itself as one of the most haunted buildings in the world. Every Halloween, the sanatorium is transformed into an enormous haunted house, dimly lit and elaborately decorated and filled with actors dressed as monsters, ghosts, and killers. S. M. and the researchers made their visit while the haunted house was in full swing.

Their group was paired with five other women who were visiting at the same time. The women must have been confused by S. M.'s boldness—from the start, she charged ahead, calling out things like "This way, guys!" as she led them deeper into the building. The actors were good at their jobs: When costumed monsters and bloodied murderers leaped out of the shadows, the other members of the group screamed and jumped. But not S. M. She smiled and laughed. Once, she even reached out to poke a monster in its masked face (she was "curious," she said later) and turned the tables by scaring the actor inside the costume.

Unsurprisingly by now, the clips the team showed S. M. from scary movies produced no fear either, although she reacted to other film footage in the ways you might generally expect of people, exhibiting sadness at sad moments, disgust at gross-out scenes, and so on. In a 2011 article detailing the whole study, Feinstein, Adolphs, Damasio, and Tranel wrote, "On no occasion did S. M. exhibit fear."

But finally, two years later, Justin Feinstein found a way to make Patient S. M. feel afraid.

THE SCIENTISTS WHO STUDY EMOTION make a distinction between what they call "exteroceptive" fear and "interoceptive" fear—the fear that comes upon us from outside of ourselves, and the fear that affects us from within. Everything researchers had done to try to trigger S. M.'s seemingly nonexistent fear response had been exteroceptive, aimed at her through visual, auditory, or other external sensory triggers. Now, though, Feinstein and his colleagues decided to take a different approach.

Previous research showed that the inhalation of carbon dioxide can induce fear, and even panic attacks, in humans. And in mice, the amygdala has been shown to be involved in the detection of carbon dioxide. Feinstein hypothesized that if he exposed S. M. to a commonly used carbon dioxide inhalation test, she would display a reduced level, relative to most other people, of CO_2-provoked fear.

On the day of the test, S. M. was asked to lie back in a reclining chair and a plastic mask was placed over her nose and mouth. Then she took one big breath through the mask, inhaling air that had been mixed with 35 percent carbon dioxide, 875 times the normal amount of CO_2 in the air we typically breathe.

This is a well-established experimental tactic, and its effects usually expire within a minute. The presence of the CO_2 in our system triggers alarm bells both in our central and peripheral nervous systems. Although the single breath of CO_2-laden air does not actually affect subjects' oxygen levels, it creates an illusory sense that they need air—what researchers call "air hunger." In one quarter of participants, the experiment tends to induce feelings of deep fear and even, in some cases, a full-blown panic attack.

S. M. did not react in the muted way that Feinstein and the rest expected. Immediately after her inhalation, she began to gasp for air. Her breathing accelerated, and at the eight-second mark after her intake of CO_2, she began waving her right hand around, gesturing at the air mask. Her toes curled, her fingers flexed, and her body locked up with tension. She said "Help me!" through the plastic covering her mouth fourteen seconds into the experiment, and one of the researchers removed the mask from her face. She grabbed his hand as soon as he'd done so. "Thank you," she said, eyes wide, nostrils flaring. But for another two minutes, she fought to breathe. She gasped and gulped for air; she waved a hand toward her throat. "I can't breathe," she said.

In the end, it was nearly five minutes before she recovered, far longer than the typical panics induced by the experiment, which last for a minute or two. "S. M. had just experienced the first panic attack of her life," Feinstein wrote later. "Every experimenter in the room was shocked. S. M. had actually felt fear. She called it the worst fear she had ever felt. In all likelihood, it was probably the first time she had experienced fear since childhood."

(Feinstein told me that he and his colleagues were "walking around that week like deer in headlights." But despite their shock, "it was a great moment," he said. "Because that's the whole point of science, to be proven wrong." You learn more when your hypotheses are convincingly overturned.)

To see if the effect was replicable, the team in Iowa reached out to a researcher who'd been studying German twins who also had extensive bilateral amygdala lesions resulting from their Urbach-Wiethe disease. The two women flew across the Atlantic and, like S. M. before them, breathed in the gas. The researchers also brought in a group of control subjects with their amygdalae intact.

Feinstein and company went three for three: Both twins experienced full-blown panic attacks when the gas hit their systems. The rate of panic among the control group, meanwhile, was much lower. Not only were the supposedly fearless patients experiencing fear, they were doing so at a higher rate, and with a greater intensity, than people without lesions.

There were two takeaways from these results. First, the presence of a functioning amygdala was not, as might have been assumed from the existing body of research, a prerequisite either for the bodily or the emotional experience of fear. And second, it seemed possible that the amygdala might have not only a triggering role in our fear responses but also a muting, suppressing, or control function. That would help to explain the exaggerated responses of the Urbach-Wiethe patients to the gas: Once their panic responses were triggered, they lacked the brakes that their amygdalae would have provided to control themselves again. They could not, as Alex Honnold apparently can, simply will themselves to calm down.

"In one breath," Feinstein wrote, "we immediately learned that the amygdala could not be the brain's quintessential and sole 'fear center.' . . . Without a functioning amygdala, S. M. was still able to experience an intense and prolonged state of fear."

THERE'S ANOTHER WAY in which S. M. expresses something we might recognize as fear. But it's not fear for herself—it's fear for her children, or at the least, if not true fear, then certainly a deeply protective instinct that she never applies to her own safety. As a mother, it seems, she can recognize an immediate threat, and respond.

Once, when a woman S. M. described as a "six-foot-five neighbor lady" slapped her young son, S. M. rushed into the fray, shoved the woman, and wound up facing down not only the "neighbor lady" but several of the woman's family members, taking on all comers before the police arrived to break things up. Another time, when her son found a small bag of crack cocaine in the yard, S. M. took the drugs to the police and told them who she thought the dealer was. Soon after, written death threats started appearing on her doorstep, and one day a man materialized in her apartment hallway, held a handgun to her head, said, "Bam!" and then walked away. But when her boy found another baggie in the backyard, S. M. went back to the police again, hoping to make her home and her neighborhood safe for her kids.

These are actions any other mother might take—nothing out of the ordinary, really, except for the part where she felt no fear whatsoever when a gun's muzzle was rested against her skull. But paired with her complete lack of concern for her own safety, even in situations when her children were out of harm's way, it suggests something: A mother's grizzly-bear instinct is not triggered by her amygdala. Our fears can spring from multiple sources within us.

These days, S. M. has very little contact with her three grown children. But she still has a mother's fears. Once, intrigued by the seeming disconnect, a researcher asked her about her son.

"Your son is now a soldier in Afghanistan, right?" the researcher said. "Are you worried about him?"

"Yes," she said. "I am."

The researcher asked her what she was worried about.

"I am worried about him being hurt, having bad things happening to him. Someone can be holding a gun to my son right now."

"There's something interesting there," the researcher said. "You basically say that if someone held a gun to you, you wouldn't be afraid. But if someone did that to your son, then you would be afraid?"

S. M. replied with a denial. "I am not afraid," she said. "I just don't want that for him. What you need to understand is that I am worried, but not afraid."

So what, then, was the difference between fear and worry?

"'Afraid' means being frightened," S. M. said. "Being scared. And 'worried' means not wanting something to happen. I have always been worried about things, but I am never afraid. If I could stand between my son and the bullet, I would do that because I am not afraid."

I WAS FASCINATED by S. M.'s existence. I've sometimes felt as though my life is less a pursuit of happiness and more an ongoing, endless duel with fear. So the idea that there was someone out there who hardly even knew what it was to feel it? I was hooked. Trying not to seem like a voyeur, I asked Ralph Adolphs what S. M. was really *like*.

"Certainly, if you just met her, and you didn't know anything, and you just interacted with her, you would be hard pressed to really find anything all that unusual," he told me. "Unless you really asked her, or you took her on a roller-coaster ride or to a haunted house or something . . . she would seem, you know, relatively normal. Like a very pleasant, friendly kind of person, but not way out in terms of her behavior." That wasn't at all unusual for lesion patients, he told me. Even amnesic patients, like the famous Patient H. M., who was almost entirely incapable of forming new long-term memories, and lived his life in roughly thirty-second increments, generally come across in casual conversation as though there is nothing out of the ordinary about their (actually extraordinary) minds.

"You're never going to ask, you know, what year is it, or who's president, or something weird," Adolphs said of his amnesic research subjects. "If you just encountered them on the street, you would ask them, 'How's it going?' And they would say, 'Fine. I'm having a great day.' You could ask, 'So what have you been doing?' 'Oh, you know, various things.'"

Adolphs noted his patients' incredible capacity to compensate, socially and otherwise, for their deficits. "They're able to construct an intact person," he told me, and that ability is increasingly a focus of the research in his lab. Rather than asking, "What is a deficit in these people?" he says to ask, "What's the rest of the brain doing to make up for what's missing?" Adolphs is continually impressed by the brain's plasticity, its ability to regroup and reorganize itself as required. "We don't know what the limits of that are, to be honest. It seems remarkable."

It *is* remarkable. Patient S. M. is now in her mid-fifties, and her health problems are beginning to pile up. But it's incredible that she has survived this long without access to humanity's primary internal alarm system. Her lesions are unique among known human cases, but here's a grim point of comparison from our primate relatives. In 1968, when S. M. was still a toddler with her amygdala intact, a psychiatrist named Arthur Kling captured a group of wild rhesus monkeys on a small island off the coast of Puerto Rico. He removed their amygdalae and set the monkeys free. Within two weeks, all of them were dead, either by starvation, drowning, or attacks from their intact peers.

When I first set out to learn about S. M., I had expected to find a cautionary tale. And yes, her life has been harrowing; her fearlessness has led her into danger, has isolated her from others, and has made her world smaller in many ways.

People, mostly men, have taken advantage of her and abused her. She has spent most of her life subsisting off government disability payments. She has often been hungry, because she isn't driven to eat, and she's not great with money—without fear of consequences, why would you be? She has difficulty maintaining friendships, apparently because she lacks the inhibitions that most of us labor under. When she meets someone she likes, she comes on strong, offers limitless generosity and asks for the same in return, and her intensity can drive people away. Turns out the fear of rejection can be a useful social handrail.

"It is a hard life," Justin Feinstein told me. "She lacks a social bubble. Typically we all have some imaginary bubble drawn around our bodies, and if anyone invades that personal space, we feel some sense of discomfort, right? Whereas with her, you could be standing literally nose to nose, tip to tip, with a stranger, and she doesn't feel at all uncomfortable." He's interested in the idea of the amygdala not only as a triggering mechanism but as a kind of brake on our behavior as well. "What you see in S. M. is really a manifestation of somebody living life without those brakes."

Life without brakes: hard to fathom for someone like me, fighting to free myself from a heavy foot on the brake pedal.

Despite everything, I found much to admire in S. M.'s life. Her trust in others, even if occasionally misplaced, is the kind of thing we could probably all use a little more of. Her adaptability, her capacity to survive despite the blank spots in the map of her brain, amazed me. I thought about her charging ahead through that haunted house, laughing, calling for the group to follow her lead. Her fearlessness made her so open to the world. I envied that, at least a little.

If the "fearless" people I'd read about were like a buffet of characteristics to choose from, I suppose that's what I'd select for myself: S. M.'s incredible openness, her boldness, and Alex Honnold's seemingly limitless capacity for calm.

In the end, though, I knew that fear was necessary. Mine might sometimes feel like it was on overdrive, but it was there for a reason: to help me survive. Justin Feinstein had described fear to me as a "key ingredient" in the continued existence of the species across millennia (and not just our species, if we define "fear" more broadly). Even if it sometimes inconvenienced me, it wasn't something to wish away.

9

why fear matters

There were four of us in the park by the Ottawa River that night. We had taken the bus downtown after school, walked across the long bridge that divided Ontario from Quebec, and bought a couple packs of Mike's Hard Lemonade from a gas station convenience store on the Quebec side. Alcohol was easy to come by there, no matter that we were not yet of drinking age. Across the river, the copper rooftops of the Parliament Buildings gleamed in the summer sun.

With our backpacks clinking, we left the busy intersection at the bridge and followed a paved pedestrian path deep into the thin strip of parkland that bordered the river. We settled onto the grass with the river in front of us and the fenced-in compound of a paper-processing plant looming behind. I'd had my first kiss in this same stretch of green space the summer before, with the Canada Day fireworks exploding overhead. As we made ourselves comfortable, I looked over my shoulder toward the paper plant and noticed a man leaning against the chain-link factory fence, watching us.

Hours passed, the sun vanished, and our bottles emptied themselves. S. disappeared into the bushes to pee; when she came back, she leaned in close, giggling, boozy, and whispered, "Guys, there's a man watching us."

I froze. I looked over my shoulder again. It was the same man.

My buzz evaporated, leaving only muddled fear; the alcohol curdled in my gut. I don't remember any discussion about what we did next—we just moved together, quickly, acting on a shared instinct. We packed up and moved back down the path in a tight cluster, herd animals in defensive formation. The wide, dark river was on our right; to our left, maybe sixty feet away in the night, we could see the man start to walk his side of the fence, alongside us, keeping pace. Ahead, a gate appeared in the fence, with a chain and a padlock looped through it. It was locked, right? It had to be locked. As we passed the gate, the man reached it, grasped its bars, and rattled the chain. We walked faster, and he kept pace—an eerie, silent chase.

"You should run for help," D. said to me. "You're the fastest."

"I'm not leaving you guys." I tried to sound valiant. The dark, tree-lined path ahead of me was monstrous. I wasn't braving it alone.

"There are four of us and one of him," S. said, like a mantra. "There are four of us . . ."

"That won't matter if he has a gun," A. whispered. We all understood that if a barrel was pointed at any one of us, the remaining three would do whatever we were told.

Without slowing, we reached into our backpacks and passed some of our collected empties among ourselves, gripping the clear glass bottles by their necks. Now each of us had a weapon.

The trees had thickened along the fence, and we could no longer make out the man shadowing us. Ahead, we could see another gate. It was forty feet, thirty feet, twenty-five feet away, and now we could see that there was no chain or padlock on this one. We didn't wait to find out if the man would reappear there—as we drew even with it, we ran. We ran and ran without stopping, without looking back, until we were back at the intersection, back on the bridge, back in the glow of street lamps and headlights pushing back the night.

THERE WAS A LESSON in the way we were driven to move that night, driven to act, but I didn't understand it yet. Twelve years later, not yet thirty years old and still newish to Whitehorse, I experienced that drive again. At three in the morning one cold winter night, I crouched behind a Toyota Matrix in the dark of a dealership parking lot and watched a streetlight glint off the roof of a taxi idling at the corner. The cabbie had been following me for five blocks.

At first, when his headlights had slowed coming toward me, I'd thought he was just looking for a fare. Walking home alone from a friend's house at that hour, in the middle of a Yukon winter, I probably seemed a likely client. But I'd waved him off, and then he'd driven past me and pulled a U-turn and shadowed me at low speed down the street. Nervous, I'd turned down another street; he followed. I started making excuses for him then—wondered if, somehow, he thought he was doing the right thing, unknowingly terrorizing me as he attempted to shepherd me safely home. But a louder voice in my head said that made no sense, something was off here, something was wrong. And I'd listened to that voice. As I'd done years earlier, I'd heeded instinct. I'd ducked into the dealership, and now I crouched in the snow, sickened by the irony: The "safe" thing to do, instead of walking the fifteen minutes home, would have been to call a cab.

The dealership was on a corner; the fenced lot had entrances on two sides of the block. The cabbie idled on Sixth, where I'd entered the maze of parked cars for sale, so I shuffled toward the Main Street entrance, sliding between Tundras and 4Runners, trying to stay hidden—feeling ridiculous, but not ridiculous enough to show myself again.

When I reached the edge of the lot, I took a deep breath and burst out onto the street and sprinted toward the lights of a hotel half a block away. Behind me, I heard the cab pull around the corner, felt his headlights spill across my shoulders as I ran.

I made it to the hotel's double doors just as he pulled to a stop at the curb behind me. I hit the doors and bounced off. They were locked.

The cab's front bumper was just the width of the sidewalk away from me. I turned, put my back to the useless hotel doors, sucked in a big breath, and screamed, "*Stop following me!*" I saw the whites of the driver's eyes as they widened above the wheel.

Just then, two grey-bearded men, walking with the uncertainty that comes from a few too many drinks, appeared out of the darkness down the block. "Hey," one called through the night, "what's going on?"

The cabbie put the van into reverse, turned, and drove away, taillights vanishing.

The two men came toward me cautiously. "I'm OK," I said. I sat down on the smokers' bench outside the hotel doors to catch my breath while they hovered at a polite distance, worrying. They offered to call me a taxi; they offered to walk me home. I thanked them and declined. I was two blocks from my apartment.

When I was ready, I took a deep breath, straightened my shoulders, and walked out of the pool of light at the hotel entrance, back into the dark.

THOSE TWO NIGHTS weren't the only times in my life that I ran as hard as I could for safety, a young woman afraid and racing away from a strange man or strange men. But they stand out from the rest. I think it's because of their ambiguity, their silence. The handful of other times that I ran, the men I was running from had said something to me that suggested I should: a shouted demand, a leering slur. These two times, though, I had made my decision based entirely on wordless clues. And despite that, I had been almost completely sure that running was the right thing to do.

In *The Gift of Fear*, security consultant and protector-to-the-stars Gavin de Becker writes about the power of intuition—the power of being afraid. De Becker is not a scientist, but he has spent decades working with people who were being stalked, harassed, threatened, and abused. He has extracted patterns from all his experience, and the book, packed with harrowing anecdotes, is the result.

It opens with the story of Kelly, a young woman who met a polite young man in the stairwell of her apartment building and reluctantly, despite her vague, unexplainable misgivings, accepted his help with carrying her groceries. Once inside her apartment, the man produced a gun, threatened her, and raped her. When he left the room, heading to her kitchen, she was overcome with the urge to move. She fled the apartment while his back was turned.

"She later described a fear so complete that it replaced every feeling in her body," de Becker writes.

> Like an animal hiding inside her, it opened to its full size and stood up using the muscles in her legs. "I had nothing to do with it," she explained. "I was a passenger moving down that hallway." What she experienced was real fear, not like when we are startled, not like the fear we feel at a movie, or the fear of public speaking. This fear is the powerful ally that says, "Do what I tell you to do." Sometimes, it tells a person to play dead, or to stop breathing, or to run or scream or fight, but to Kelly it said, "Just be quiet and don't doubt me and I'll get you out of here."

De Becker's thesis is that we know more than we realize about the threats the world poses. We have the power, he argues, to correctly determine when we are at risk and when we are not—we just have to learn to listen to our instincts instead of drowning them out with politeness, with thoughts of what's expected of us, with social norms. "Like every creature," he writes, "you can know when you are in the presence of danger. You have the gift of a brilliant internal guardian that stands ready to warn you of hazards and guide you through risky situations."

It's a powerful idea—an empowering idea. And in the two decades since his book was published, science has helped to fill out some of the gaps in our understanding of what de Becker generally calls intuition.

The second chapter of *The Gift of Fear* begins with another anecdote, one that hints at the mechanisms of intuition, how it works, how it warns us. Robert Thompson, a commercial pilot, tells the story of a night when he walked into a convenience store, planning to buy some magazines, and then felt suddenly, inexplicably afraid. Thompson turned around and walked out again with his magazines unpurchased.

The next man to walk into that same store, moments later, wasn't so lucky. He was a police officer, and his appearance startled the man who was in the midst of holding up the store clerk at gunpoint. The officer was shot and killed.

"I don't know what told me to leave," Thompson told de Becker. "It was just a gut feeling."

He paused, reassessing his own words. "Well, now that I think back, the guy behind the counter looked at me with a very rapid glance, just jerked his head toward me for an instant, and I guess I'm used to the clerk sizing you up when you walk in, but he was intently looking at another customer, and that must have seemed odd to me. I must have seen that he was concerned."

Later, Thompson remembered that the customer was wearing a heavy jacket on a hot night, and that there'd been a car idling in the parking lot. In de Becker's thesis, all those details added up, unconsciously, in "a cognitive process, faster than we recognize and far different from that familiar step-by-step thinking we rely on so willingly."

"Intuition is soaring flight compared to the plodding of logic," he writes.

We now know that intuition has some tangible help—like, for

instance, our ability to smell fear. When I first read about Robert Thompson's near miss, I thought immediately of Lilianne Mujica-Parodi, the Stony Brook researcher who proved the existence of human alarm pheromones by having her subjects jump out of planes. I had to wonder: Did Thompson's senses pick up on the fear scent rolling off the clerk, and likely off the robber, too? Did that cause his amygdala to trip his internal alarm systems, heightening his attention to the widened eyes and frightened face of the clerk? I have to figure that they probably did.

Mujica-Parodi wasn't interested in human alarm pheromones purely for their own sake. Neat as alarm pheromones are, she had bigger-picture questions in mind. It turns out that her skydive-based research came about as part of a wider search for answers about how our fear can serve us, how it can be a tool rather than a burden.

Mujica-Parodi had become interested in studying fear responses in healthy individuals rather than in the far more commonly studied groups, people suffering from an excess of fear and anxiety. With funding from the US military (because, after all, it has an interest in finding out how its personnel will respond to scary situations), she started to grapple with a fascinating question: What kind of person would make the best Navy SEAL?

"It was kind of a broad problem that allowed me to think more deeply about what individual variation looks like," Mujica-Parodi told me. "What is it that makes, let's say, a Navy SEAL different than a normal healthy person? And then what makes them different than someone who is more anxious? And then, going further, than people who are pathologically anxious?"

The work on fear pheromones was part of her search for an answer to that broader question. So was a second, more recent experiment during which, once again, Mujica-Parodi sent a pack of first-time skydivers up in a plane. This group, though, was different; she selected her research subjects only from people who had already signed up to go skydiving of their own volition. The first batch had been selected at random; they weren't necessarily people who *wanted* to jump out of a plane (and so, I suppose, they were more likely to produce plenty of fear-sweat). These later jumpers were a different breed, a group that scientists refer to as high-sensation seekers, or HSSs.

Mujica-Parodi and her colleagues wanted to know if it was possible to distinguish, physiologically, between two types of people who they labeled either "brave" or "reckless"—meaning, between people who took risks knowingly and deliberately, with a full understanding of the threat, and people who did not fully grasp the risks they were taking or the threats they faced. They hypothesized that the difference between bravery and recklessness was not merely a societal judgment rendered after the fact, based on the consequences of the risks taken, but that bravery and recklessness involved qualitatively different approaches to risk, differences that were "neurobiologically, physiologically, and cognitively distinct."

They started by selecting thirty aspiring skydivers, and over the course of two strictly regimented days, those subjects underwent an array of tests. They filled out standardized questionnaires about their risk avoidance behaviors (or lack thereof); they had their endorphin and adrenalin levels tested; they had their amygdala activity measured by fMRI. During the plane ride up to jumping altitude, with their senses heightened by anticipation, the subjects completed a task in which they had to rapidly identify partially obscured images of human faces as being "neutral" or "aggressive."

The results supported Mujica-Parodi's hypothesis. The "reckless" skydivers were the least attentive to risk or threat, and in some ways their brain activity had more in common with anxious people, their opposites in threat assessment, than with the more optimally balanced "brave" subjects. The implications for her research were enormous. "The conclusion that I came to after about twelve years of research is that I had sort of a category error," she told me. "I had been thinking about it in the wrong way."

Before, Mujica-Parodi (and the military) had been conceptualizing the problem in a certain way: placing people on a spectrum of stress resilience, from more resilient to less resilient. Presumably, they had figured, the best Navy SEAL would be someone who was maximally resilient to fear-inducing and stressful situations—someone as close to fearless as ordinary folks (rather than the Alex Honnolds or the Patient S. M.s of the world) can get. But actually, when she dug in, Mujica-Parodi realized that what mattered most wasn't the subject's resilience level at all. Instead, she came to view threat detection, a different function entirely, as the key.

"The distinction between different types of people doesn't become evident when you put a gun to someone's head," she told me. "If there's an actual threat, everybody reacts essentially the same way. What distinguishes people is how they respond to a potential threat. That is, an ambiguous threat." People who are generally more anxious can tend to see threats where they don't truly exist; conversely, people on the opposite, reckless, end of the anxiety spectrum can sometimes ignore or disregard genuine threats. "And that's where the kind of stress-resilience way of thinking about things breaks down," said Mujica-Parodi. "Because the ideal Navy SEAL is not somebody who is fearless. . . . Ideally, you want someone who is very good at identifying threat but doesn't identify threat where it doesn't exist."

Here's the ideal, not just for a Navy SEAL, I suppose, but for anyone who wants to successfully navigate their fears: someone who can correctly identify a threat, neither under- or overestimating its risks, and then override their initial fear response—a freezing instinct, say, which can be useful if you're a mouse trying to evade an owl in a nighttime field, but not if you're a person on a highway who's about to be hit by a U-Haul truck—in order to react effectively to mitigate said threat.

Seems clear enough when you break it down, right? But most of us will spend our lives failing at the task in one way or another, underreacting or overreacting, and hopefully living to try again the next time. That's one reason why fear memories are so powerful, why they can hang around and hurt and haunt us: They are designed to be potent, to last, to serve as shorthand when the same threat crops up again. Fear memories can enable a rapid response. Their job is too important to be easily short-circuited.

I GUESS I'VE BEEN pretty lucky. Just once in my life, I have called the police because I believed I was in danger of attack by another human being. It was a few years ago, maybe a year or two after the night that I ran from the cab driver.

At first, I thought I was just dealing with a creep. I was dog-sitting for a friend, living in her house and taking care of her husky. Early one Saturday morning, the phone—the house land line—rang at 5:00 AM. Thinking there might be some emergency, I struggled out of bed, half-awake, and answered.

"Hello?"

"Heyyyyy," came the long-drawn-out syllable. A man's voice.

"Hi," I said, uncertain.

There was a pause. Then: "I'm touching my dick."

I hung up. Seconds later, the phone rang, and I knew it must be the same man. I unplugged the handset from the wall. Somewhere else in the house, a phone rang again, and again. I found my way through the darkness to this second phone, unplugged it too as it rang and rang, and then the house was quiet. I was unsettled, but I didn't feel unsafe. Still, it took me awhile to fall back to sleep.

Two weeks later, I was back at home, in my own apartment. It was a Saturday morning, at 5:00 AM. The phone—my land line—rang. I had forgotten about the last call, and again I struggled, half awake, to get to the phone on my desk in the living room.

"Hello?" I said.

"Heyyyyy," said that same soft voice, and, with a rush of recognition, I hung up the phone before he could get another word out. Then I moved very quickly, unhesitating, with instincts I didn't really know I had, taking actions I can only figure I must have drawn from years of exposure to crime novels and bad television. I unplugged the phone from the wall. I turned off the lights I'd flipped on during my trip from bed to desk, so no one outside could see any movement in my apartment. I checked the lock on the door, and I grabbed my cell phone and the largest knife in my kitchen. Then, having prepared as best I could, I retreated to my bedroom, which had only one small window, and shut the door behind me.

I sat down cross-legged on the bed, and now that I was no longer in motion, I felt my fear, really felt it, for the first time. My heart pounded. My eyes were wide, my chest tight, my breathing short. I was terrified.

Knife in hand, I called the police from my cell. "I think someone is stalking me," I told the dispatcher.

In the moments after I heard the man's voice on the phone, I'd made a series of rapid calculations. First he had called me at my friend's house; now he had called me at home. Same man, same time of day, same day of the week. Two different land lines. There was only one thing tying the two calls together, and that was my physical presence. My conclusion, then, delivered in an instant, was that the caller knew both who I was and where I was. He had to have been physically following me, I figured, and learning enough about me along the way to track down the phone numbers at both places.

I didn't get to sleep again that night, or rather morning, and later that day I delivered a grainy-eyed report to a kind and sympathetic officer at the police station. He said he'd look into the phone records and get back to me. Then, by coincidence, I left for two weeks on a work trip. I was glad to be leaving my apartment, which now felt compromised, behind.

Years later, reading Gavin de Becker's book, I thought about those few seconds after I'd hung up on the man. There are so many times in my life when I've been afraid, in one way or another, but not many when I've been driven by fear to that kind of fast action, the rapid conclusion and immediate response. It reminded me of the night my friends and I had made our unanimous departure from the park, all those years before—our instant, silent agreement that the threat was real and immediate.

In that case, we never found out if we were right to view the man by the river as a threat, if we had made an accurate assessment. We would never know if he was what we'd all feared he might be, the words we never spoke out loud to each other as we fled: a rapist and murderer, an episode of *Law & Order: SVU* come to life.

But I did find out more about my "stalker"—and as it turned out, my conclusion was wrong.

Two weeks later, when I got back from my trip, the Mounties (Canada's federal police force, the Royal Canadian Mounted Police) told me that they'd gotten a number of reports from other women about vulgar, harassing phone calls on the same dates that I'd received mine. On the morning that I was at my friend's house—and my friend's last name happened to begin with *M*—several other women listed in the phone book under *M* got calls, too. On the morning I was at home, in my own apartment, he'd hit the *H* section of the book. The other women had been grossed out or unsettled enough to report the calls, but since I was the only one who—through sheer alphabetical bad luck—got two separate calls, at two separate locations, I was also the only one who'd become convinced I was being watched and that my life could be in danger.

I felt silly. And my embarrassment was made worse when the local paper ran a short item about the harassing phone calls. Within hours, some of my fellow Whitehorse residents posted comments online, below the news story, mocking anyone who would be frightened by a pesky phone call or two. Had I overreacted? Was I wrong to be afraid?

At the time, I felt angry at the people who'd mocked my fear, but their contempt also made me doubt myself. More recently, I found some validation in de Becker's book:

> Not everything we predict will come to pass, but since intuition is always in response to something, rather than making a fast effort to explain it away or deny the possible hazard, we are wiser (and more true to nature) if we make an effort to identify the hazard, if it exists.
>
> If there's no hazard, we have lost nothing and have added a new distinction to our intuition. . . . Intuition is always learning, and though it may occasionally send a signal that turns out to be less than urgent, everything it communicates to you is meaningful.

I thought about the man on the phone. Was he calling from outside my apartment that night, ready to kill me? No, he wasn't. But was he someone who enjoyed frightening women, who found some sexual gratification, perhaps, in frightening us, in displaying his power to make us afraid? It seemed likely. And would I ever want to be alone in a closed room with him? No, almost certainly not. He was a person to avoid—a person, at least on some level, to be afraid of. I had been right about that much.

The catch is in figuring out when to listen to our fear—when to trust our threat assessment, which may or may not be as accurate as the ideal Navy SEAL's—and when to suppress or ignore it. And I had a lifetime of reasons not to trust my own reactions.

When I called Justin Feinstein to ask about Patient S. M., we also talked about the broader role of fear and its appropriateness or inappropriateness in any given situation. In modern society, we are facing, Feinstein said, a "juxtaposition right now between fear as an emotion for survival, and fear as an emotion to the detriment of our survival."

"It really affects us not just as individuals," he said, but "at a societal level." Our fear has been a survival tool for as long as we have existed—and even before we existed as *Homo sapiens*, we had access to less advanced forms of threat response than the complex, multivalent ones we now call fear. But in today's world, our ancient alarm systems seem less and less in touch with our contemporary dangers. (While my internal alarm system shrieks about heights, there is nothing warning me to get up off the couch and stop wallowing glumly in other people's choreographed happiness on Facebook.)

"The usefulness of fear as an emotion to help us survive in the wild is starting to be tested by modern society," Feinstein said. "There's a paradox happening now, where as a society we have everything that we could potentially want to imbue ourselves with—safety, and certainty, comfort, things that our ancestors couldn't have dreamed of—yet our anxiety and fear is off the charts." It's that paradox, he explained, that had driven him to study fear in the first place. "We need to study fear, we need to study its purpose, and we need to study how it's actually leading us in a potentially maladaptive direction."

After I read *The Gift of Fear*, I spent some grim late nights at home, thinking back on the times in my life when I've been afraid of another person, analyzing each one, trying to take them apart. How had I known to act, if I did act? Had I been right? How had I acquired the tools, over the years, to make those snap judgments?

I suppose, when my phone rang at 5:00 AM in my apartment, I relied not only on the data from the call two weeks earlier. All the information I'd collected during a lifetime of reading the news, watching TV, and talking to female friends and acquaintances about creeps, stalkers, and predators was fed through my mental machinery again as my brain calculated the threat posed by the man on the phone. I acted on the conclusion that came shooting out the other end of the machine.

I was right to be alarmed by the creepy caller—everything in my experience told me so, even if I misjudged the scale of the threat. But it's easy to imagine situations where people's fear responses are completely wrong, and dangerously so. "If there's no hazard, we have lost nothing," Gavin de Becker wrote. But what if our actions in response to a perceived threat *have* caused loss? What if our fears have caused us to lash out? What if, say, our misapplied fears have led us to call the police on a black boy playing with a toy gun, and then a police officer's irrational fears, in turn, have led them to shoot to kill, without pausing to assess the situation?

Rifling through the drawer of scary experiences in my memories, I wanted to embrace Gavin de Becker's theory. I wanted to believe that, as he wrote, "If your intuition is informed accurately, the danger signal will sound when it should. If you come to trust this fact, you'll not only be safer, but it will be possible to live life nearly free of fear."

I couldn't do as de Becker advised and set aside my reluctance to trust my own fear responses so completely. Fear can be a useful tool for survival, yes. But I worried about the fallout of fear gone awry—fear that had become "maladaptive," to use the term preferred by the neuroscientists I'd spoken to. Hadn't I seen for myself on the Usual that an outbreak of wild, irrational fear could put not just a fearful person but also the people around her in danger?

So how can we know when to trust our fear?

In the end, I can only fall back on the clarity I've felt in the times I was driven to action. My irrational fears—my belief that I would slide to my death down the Duomo's terra-cotta roof tiles, or that the wind would blow me off a hiking trail, or that my car might plummet off any curve of a wet highway—have always been paralyzing. They've made my brain fuzzy, my movements slow and awkward. But in the handful of times in my life when my fear has propelled me into rapid, instinctive movement, I've felt different: not fuzzy at all, but sharp—as sharp as the edges of torn metal on the side panels of my Jeep after I'd steered myself away from an almost certainly fatal impact.

"I have ninja skills," I'd told Svenja, as I sat on the couch in her office, pods buzzing rhythmically, my eyes shifting back and forth, back and forth, behind my eyelids. "I saved myself." I had believed it when I said it. And now, I thought, maybe I could believe in that sharpness—could try to remember how different it felt from the fuzzy, paralyzing fear. That, at least, was something to hold on to.

One summer day during my undergrad years, a friend of mine went for a bike ride on a path near her home. While she was riding, a man on a bike stopped her to ask for directions. She explained where the path went, and when he then invited her to join him on his ride, she turned him down. She felt strange about the interaction, though she couldn't say exactly why. Something about the way he made eye contact seemed off, she felt. The man had a weird vibe.

After she declined his invitation to carry on together, she turned her bike around and rode back in the direction she'd come from. She didn't hear him coming until he was almost on top of her, but suddenly he was there, on her tail, riding hard. He was close enough to reach out and touch her. On instinct, she swore at him and rode away fast. He fell back and didn't continue the chase.

She started doubting herself almost immediately after he was gone. She wondered if she'd misunderstood him somehow, if she'd read aggression where there had only been an awkward attempt at playfulness. Or . . . something?

As she rode away, she yelled back into the empty woods, "I'm sorry, you just scared me!"

She didn't think much about the man again until she saw the missing persons announcement. Another young woman had disappeared that same morning, on the same bike path. Search parties were being sent out. They found her body in the woods near the path a few days later. Eventually, the man on the bicycle was convicted of her murder.

Years later, when we talked about that day again, my friend told me she'd never really experienced any lingering trauma or sense of danger from her close call. Sadness and anger, yes—but not fear of the world. She thought it was because she'd had agency, because she had sensed a threat and taken effective action. She hadn't been trapped helplessly with her fear, like Pavlov's dogs in the flood.

Her moment of sudden, immediate fear had kept her alive.

epilogue: a détente with fear

One month after my visit to Amsterdam, I went to Moab, Utah, for a camping trip with my cousin Nathan and his family. Moab is a small town in a vast sandstone desert, the adventure capital of southern Utah. It draws hikers, rock climbers, whitewater rafters, mountain bikers, off-roaders, and more to play in its surrounding canyons and mesas. Where better, I thought, to try to test drive my cure?

I had given a lot of thought to the testing process. The catch was this: Some fear of heights is natural and healthy, and I didn't want or expect a complete excision of those sensations. What I had hoped to get from Merel Kindt was relief from my over-the-top, irrational reactions. That meant, I figured, that the test had to be fundamentally safe, targeting only my unreasonable fears. A friend had suggested that I head to the Grand Canyon, a half-day's drive away, and stand right on the edge and look down. But I knew that a stumble there really would kill me (as if to emphasize the point, a tourist had fallen into the canyon just the week before), and so I didn't expect to find any comfort in the view from the top all the way down to the bottom.

Eventually, cruising the websites of various Moab tour operators, I settled on a zip-line operation. Zip-lining, I figured, would have me dangling high in the air, voluntarily stepping from safety into open space, and flying rapidly downhill—some of my least favorite things. It was exactly the kind of activity I would have avoided before my cure, not out of any genuine fear that I might come to harm but in order to preempt the possibility of making a humiliating scene. It was perfect.

I booked online, and when the time came, I left my family at the campground—where Nathan's toddler was riding his balance bike furiously, fearlessly around our quiet loop road, making me wonder if I'd ever been that bold—and headed to an office just off the highway on the far side of town. I guiltily signed a waiver promising that I had no physical or mental health conditions that would preclude my participation in the day's itinerary, and thought, *This better work.*

After a safety briefing, my group cinched our harnesses and piled into two large ATVs for a wild ride up the rubbled sandstone cliffs behind the company's office building. I white-knuckled the grab bar as we climbed up a seemingly impossible trail, and the kids behind me shrieked with that familiar mixture of terror and delight—the sound born of haunted houses, of roller coasters, of a cold sprinkler on a hot summer day.

At the top of the cliffs, we left the vehicles behind and hiked up a sandstone fin to the first zip line. From here, there was no changing my mind, no easy escape beyond finishing the six-line circuit with the rest of the group. Nervous, unsure how my body would react, not wanting to give myself time to stew, I volunteered to go second. A guide checked my harness, clipped me in, and waited to hear through his walkie-talkie that the other guide, who'd already zipped over to the far end of the first line, was ready to receive me.

When I got the all-clear, I sat down in my harness, took a deep breath, lifted my feet free of the rock, and let gravity take me. My stomach churned as I glided into space, picking up speed, but the nerves faded as I made that first crossing. I could, I realized, look around at the scenery without distress; I could look down at the ground flowing far below me. My chest was loose, my breathing free. My body wasn't responding to a perceived threat. I wasn't panicking, crying, freezing up, embarrassing myself, outing myself as a lying, fraudulent waiver-signer.

When I reached the far side, I asked Nate, the guide, as he unclipped me, "Do they get a lot scarier after this one?" I knew that some lines were longer, faster, higher. I worried that the worst was yet to come. I tried to sound casual, like I was just making small talk.

"Nothing about this scares me," he said, unhelpfully. And then, "But if we got you across the first one, you'll be fine." I would be fine!

Nate was right. With each zip line, the chasms I soared over got deeper, the rides longer and faster, but I remained calm—calmer than I'd been on that first, nervous ride. I looked down at the rocky gullies below me as I buzzed along, I looked up and around. On one line, Nate dared me to take a backward start, and I did it: I stepped off the rocks and into space with my back turned to the drop and swooped away.

Later, feeling more confident, I asked the other guide if anyone ever signed up and then freaked out and became unable to complete the course. He said he'd never had anyone refuse to go on once they'd started, but they did get the odd customer who arrived at the start and then decided they couldn't or wouldn't go through with it. Earlier this season, he said, they'd had one woman who wept through the entire circuit—but she made it to the end.

That could have been me, I realized. Or worse. I sent a silent thank-you to the universe, and to Dr. Kindt's team in particular, that I hadn't become the tour's latest Sobbing Customer.

After the last customer had ripped across the last zip line, we piled back into the ATVs for the ride down. I was worried, still, a little, about this part: The ride up had been perilously steep, and I wondered how my body would react to the downhill view as we bumped and rocked our way back to town. I opted to sit in the back seat so I wouldn't be staring through the windshield, hoping that might obscure my sense of our downward plunge.

But I didn't need to worry, I realized, as we teetered down the trail, the vehicle swaying. I felt OK. Occasionally, as we rounded a bend and a new vista appeared below us, I got that little swoopy feeling in my stomach, but that was fine and natural, I supposed. The two young girls on the tour, sitting in the front seat this time, screamed and screamed with delight, and I tried to let their joy infect me, tried to embrace the swoopy feeling instead of fearing it. It worked: By the time we reached the base of the cliffs, I was laughing.

epilogue

Unlike many people, I've never really enjoyed the feeling of being afraid. Fear for me has rarely been a thrill, a sensation to seek out and ride its wild crest. Instead, I've experienced fear as a force that limited me, that made my world smaller.

Growing up, I never liked haunted houses, or anything else that might involve a deliberate effort to scare me—even if it was meant to be fun, to provoke the kind of shriek that dissolves into laughter. That aversion predated my epilepsy diagnosis, but once I associated nightmares with seizures, it only got stronger. I avoided scary movies, and read frightening books with caution. (On an episode of the TV show *Friends*, I had seen Joey store his copy of *The Shining* in the freezer—the act, he claimed, kept him "safer," even if not perfectly safe. When I read Stephen King's *It* in junior high, a rare-for-me foray into the horror genre, I took a cue from Joey and did the same.)

But now, years later, I thought about those young girls screaming with joy and fear, the emotions intermingled, in the ATV on the way down the red sandstone cliff above Moab. I thought about one of the other zip-lining customers, a woman from California, who whooped and shrieked her way along each line of the circuit, shouting her delight across the canyons. I thought about the people who scientists call "high-sensation seekers," the people who chase the intense thrills of recreational danger, like Kelsey, and the two Codys, and everyone else at the skydiving camp.

And I thought about Patient S. M., who was not, after all, indifferent to fearful experiences. It wasn't just that they didn't scare her, leaving a void where the feeling should be; she enjoyed them, even sought them out. She was excited by the snakes and the spiders, the monsters jumping out of dark corners in the haunted old sanatorium building. (Feinstein and the rest speculated that the only thing holding her back from becoming a classic high-sensation seeker and engaging in activities like skydiving was her lack of disposable income.) The things that thrilled and frightened other people were pure fun for her—there was, after all, nothing wrong with her brain circuitry for enjoyment. She may have lacked the "brakes" that a functioning amygdala provides, but she had a working gas pedal.

I had never been much like those people, I'd always thought. I conceived of myself as a deeply fearful person—and the fears that had marked my life were real, and painful. But the more I thought about it, the more I realized I had sometimes found ways to enjoy a thrill or two. I had loved plunging through a rapid in a canoe, fear flickering on the edges of my concentration as I dug my paddle into the water. I had felt the joy of steering a mountain bike down a narrow dirt trail, balancing on my pedals and ducking under tree branches. I had even, now and then, managed to truly enjoy my efforts to climb on rock and ice. I had been working to stop tapping the brakes so frequently, but I, too, could press down on the accelerator.

I thought about the patterned nature, the essential circularity, of my fears. "He who fears he shall suffer," Michel de Montaigne wrote, "already suffers what he fears." He was right: I had spent so much time being afraid of . . . being afraid. With my fear of heights, each attack of paralysis and panic—rare as they were in the grand scheme of my life—had made the future possibility of another one seem more ominous. My fear of driving, too, had occurred on a loop: my memories of the earlier crashes rising up in my mind, seizing control, leaving me terrified of the past repeating itself. And my fear of my mom's death had been driven so powerfully by my awareness of her own experience of losing *her* mother. Here was a common theme: I feared the past coming to pass again.

Sometimes, thinking about fear and my aversion to it, and trying to make sense of how all the shards of my different fears fit together, I've looked back on that little girl I once was—the one who came home from school and told her mom that she never ran quite as fast as she could on the playground, for fear of losing control—and wondered, *How many of my seemingly disparate fears are really about control—about keeping traction on the unsteady surface of life?*

The loss of control was at the heart of my problems with driving: I feared, dreaded, that remembered feeling of the tires losing their grip on the road. So many of my heights-panics, too, have revolved around the idea that I might slip: that my feet might come out from under me on a frozen creek or a steep trail, that the wind might blow me head over heels, that I might lose my balance and tumble over the railing of that dome high above Florence. And my mom's living or dying was always going to be beyond my control.

I thought about agency, about my friend's inexplicable flight from an unknown danger on the bike path, about the times I'd been driven into action. I thought about my own obliviousness in the spinning SUV, how my belief that everything would be fine and I was still in control had, in the end, protected my mind from the trauma of the wreck moments later. Agency, even if only the illusion of it, seemed like one possible cure for fear.

But even an illusory sense of control is not always available to us. Maybe something else to strive for, beyond control, is acceptance—acceptance of the fact that fear happens, for good reasons and for bad. That it's OK, sometimes, to be afraid. That it can even, sometimes, be fun. These seemed like lessons to take with me into the future.

NOW, THREE YEARS after I'd begun this project of facing my fears, confronting them, trying to renegotiate my relationship with them, it was time to take stock. How had I done?

My fear of driving, the legacy of my traumatic series of car crashes, had been resolved entirely. I'd been freed of the weight of those awful memories and could enjoy the open road again. My fear of loss, the fear of my mom's death that had haunted me for years and then threatened to transform itself into a consuming fear of others' deaths, too, had been defused, to some extent at least, by my new understanding of my own resilience. There would be more grief in my future, but I was readier for it now. My dread on that front was largely gone.

Then there was my fear of heights, which offered a more lingering question mark. I suspect that, at a minimum, Merel Kindt has successfully cured my fear as it relates to dangling in space—as I did in the bucket of the fire truck, and on the zip line. Whether it will apply perfectly to a steep, exposed slope, or the mast of a tall ship, I don't yet know. My guess is that, in those contexts, my old fear won't be gone entirely.

But I realized something in the process of ransacking my memories in search of my most fearful moments. Going over chronologies, replaying scenes in my mind, I noticed that my worst panics, with the exception of the one on the Usual, were all a very long time ago. They were extremely rare, too—something that was easy to overlook when I was listing them all off. And with that one exception, they predated all my efforts to treat my own fear. I wondered if maybe I was fearing the recurrence of something that wasn't likely to come back again anyway.

Because here's the thing about the incident on the Usual. I was, as my roommate on the Arctic cruise ship had reminded me, not myself at that time. I was depleted by grief and isolation; I was a more fragile, more raw version of Eva. And the trauma of my last two car accidents, the rollovers coming one after the next during that same long winter of grief, had been compounded by my sadness and anger and loss, too.

Suddenly everything seemed connected—and, suddenly, that made a life less filled with fear seem that much more possible. If you set aside the Usual, marked it with a grief asterisk, I hadn't had a true heights-panic in nearly a decade. And in that time, I had tackled heights far more extreme than the places that had caused my long-ago meltdowns. I wondered if my fear might now be less potent than I had imagined, that it might in the future hold less power over me than I had allowed it to in the past.

Yes, it's possible, even likely, that I will still feel discomfort around exposed heights in the future. But discomfort, in comparison to frozen, life-endangering panic, is manageable; I can live with that probability. And I can live with the likelihood, too, that as I age I may find new fears. I have better tools now, better understanding. I am less afraid of fear itself.

JUST AS I NEVER KNEW, until recently, about my dad's childhood fear of heights, I don't know much about what scared my mom. The only fears I can remember her expressing were for me, or about her ability to parent me.

She wasn't generally an anxious parent—didn't hover, didn't try to keep me from being exposed to all the world's rough edges. But every now and then, she would be struck by powerful and bizarrely specific fears for my safety. She was afraid, when I went on a boozy spring break trip to Mexico at the end of high school, that I would be trampled to death in a nightclub fire; she made me give a solemn promise to look for the exits and plan my escape route upon arrival. A few years later, when I traveled to Dover during a holiday from grad school in England, she had a nightmare that I had fallen from the famous white cliffs, and in the morning she couldn't shake her dread. I got a panicked email asking me for an immediate reply.

Still, although she didn't seem like an unusually fearful person, I worried often about frightening her. I viewed her sadness as fragility—I didn't yet understand the depth of her resilience and strength. I told myself that there were things I couldn't do, paths I couldn't take, lives I couldn't live, because they would scare my mother too much. I didn't want to terrorize her. But I've wondered to what extent I was using my concern for her as a way to dodge my own fears, to live a safer, quieter, more avoidant life.

I know now that I don't need to make my world smaller—I don't have to allow fear to shrink the boundaries of the life that I live. But I also know that I don't have to keep trying, pushing, proving myself. I don't have to become a rock climber if I don't enjoy rock climbing, even if it doesn't scare me as much as it used to. I can choose to seek thrills, to embrace the rush of fear, or I can choose to stay home and read a good book. Maybe I'll go back to Florence someday; maybe I'll try to learn to sail again.

But if I don't, I'll know that it's not because fear stopped me. If I never make it back to the top of the Duomo, it'll be because there was so much else to do and see in the world. My time is not limitless—something I can now accept, mostly, without fear.

selected bibliography

Adler, Shelley R. "Sudden Unexplained Nocturnal Death Syndrome Among Hmong Immigrants: Examining the Role of the 'Nightmare.'" *Journal of American Folklore*. 1991.

Amaral, David, Ralph Adolphs, eds. *Living Without an Amygdala*. New York: Guilford Press, 2016.

Bourke, Joanna. *Fear: A Cultural History*. London: Virago, 2006.

Bourne, Edmund J. *The Anxiety and Phobia Workbook*. Oakland, CA: New Harbinger Publications, 2015.

Bridgeman, Bruce. "The power of placebos." *The American Journal of Psychology* 112, no. 3 (1999).

Coelho, Carlos M., Guy Wallis. "Deconstructing Acrophobia: Physiological and Psychological Precursors to Developing a Fear of Heights." *Depression and Anxiety* 27, no. 9 (2010).

Cover Jones Mary. "A Laboratory Study of Fear: The Case of Peter." *Pedagogical Seminary* 31 (1924).

Damasio, Antonio. *Descartes' Error*. New York: Grosset/Putnam, 1994.

— — —. *Looking for Spinoza*. Boston: Houghton Mifflin Harcourt, 2003.

De Becker, Gavin. *The Gift of Fear*. New York: Dell, 1997.

Didion, Joan. *The Year of Magical Thinking*. New York: Vintage Books, 2005.

Dittrich, Luke. *Patient H. M.* New York: Random House, 2016.

Dowling, John E. *Understanding the Brain*. New York: W. W. Norton, 2018.

Edelman, Hope. *Motherless Daughters*. Boston: Addison-Wesley Publishing Company, 1994.

Elsey, James. W. B., Merel Kindt. "Breaking boundaries: optimizing reconsolidation-based interventions for strong and old memories." *Learning & Memory* (2017).

Feinstein, Justin S. et al. "Fear and panic in humans with bilateral amygdala damage." *Nature Neuroscience* 16, no. 3 (2013).

Feinstein, Justin S., Ralph Adolphs, Antonio Damasio, Daniel Tranel. "The human amygdala and the induction and experience of fear." *Current Biology* (2011).

Gonzales, Laurence, *Deep Survival.* New York: W. W. Norton, 2004.

Honnold, Alex, David Roberts. *Alone on the Wall.* New York: W. W. Norton, 2016.

Jaycox, Lisa H., Edna B. Foa, Andrew R. Morral. "Influence of Emotional Engagement and Habituation on Exposure Therapy for PTSD." *Journal of Consulting and Clinical Psychology* 66, no. 1 (1998).

Kindt, Merel, Marieke Soeter, Bram Vervliet. "Beyond extinction: Erasing human fear responses and preventing the return of fear." *Nature Neuroscience* 12, no. 3 (2009).

Kugler, Gunter, Doreen Huppert, Erich Schneider, Thomas Brandt. "Fear of heights fixes gaze to the horizon." *Journal of Vestibular Research.* 2014.

LeDoux, Joseph. *Anxious.* Penguin, 2016.

———. *The Emotional Brain.* New York: Simon & Schuster, 1996.

———. *Synaptic Self.* New York: Penguin, 2003.

MacKinnon, J. B. "The Strange Brain of the World's Greatest Solo Climber." *Nautilus,* August 11, 2016, 039.

McClelland, Mac. *Irritable Hearts: A PTSD Love Story.* New York: Flatiron, 2015.

Metter, Julian, Larry K. Michelson. "Theoretical, clinical, research, and ethical constraints of the eye movement desensitization reprocessing technique." *Journal of Traumatic Stress* 6, no. 3 (1993).

Mujica-Parodi, Lilianne R., Helmut H. Strey, Blaise Frederick, Robert Savoy, David Cox, Yevgeny Botanov, Denis Tolkunov, Denis Rubin, Jochen Weber. "Chemosensory cues to conspecific emotional stress activate amygdala in humans." *PLoS ONE* 4, no. 7 (2009).

Mujica-Parodi, L.R., Joshua M. Carlson, Jiook Cha, Denis Rubin. "The fine line between 'brave' and 'reckless': Amygdala reactivity and regulation predict recognition of risk." *Neuroimage* 103 (2014).

Nader, Karim, Glenn E. Schafe, Joseph LeDoux. "Fear memories require protein synthesis in the amygdala for reconsolidation after retrieval." *Nature* 406 (2000).

O'Farrell, Maggie. *I Am, I Am, I Am.* Toronto: Knopf Canada, 2017.

Robb, Alice. *Why We Dream.* Boston: Houghton Mifflin Harcourt, 2018.

Rubin, Denis, Yevgeny Botanov, Greg Hajcak, Lilianne R. Mujica-Parodi. "Second-hand stress: Inhalation of stress sweat enhances neural response to neutral faces." *Social Cognitive and Affective Neuroscience* 7, no. 2 (2012).

Sacks, Oliver. *The Man Who Mistook His Wife for a Hat.* New York: Touchstone, 1998.

———. *The River of Consciousness.* New York: Knopf, 2017.

Saul, Helen. *Phobias.* New York: Arcade, 2012.

Shapiro, Francine. "Efficacy of the eye movement desensitization procedure in the treatment of traumatic memories." *Journal of Traumatic Stress* 2, no. 2 (1989).

———. *EMDR: The Breakthrough Therapy for Overcoming Anxiety, Stress, and Trauma.* New York: Basic Books, 1997.

Soeter, Marieke, Merel Kindt. "An abrupt transformation of phobic behavior after a post-retrieval amnesic agent." *Biological Psychiatry* 78, no. 12, (2015).

Van der Kolk, Bessel. *The Body Keeps the Score.* New York: Penguin Books, 2014.

Walker, Matthew. *Why We Sleep.* New York: Scribner, 2017.

Watson, John B., Rosalie Rayner. "Conditioned Emotional Reactions." *Journal of Experimental Psychology* 3, no. 1 (1920).

notes on sources

Beyond its reliance on my own memories, Nerve is built primarily on published books and some scholarly articles, most of which are listed in the selected bibliography. (In addition to reading widely, I conducted original interviews with Lilianne Mujica-Parodi, Edna Foa, Merel Kindt, Justin Feinstein, and Ralph Adolphs.) I have generally attempted to credit my sources in the text, but anyone interested in digging deeper will find some additional notes on sourcing below.

Chapter Two

I found the G. Stanley Hall quotation via Joanna Bourke's *Fear: A Cultural History.* The reference to Ares's sons, Phobos and Deimos, came from Joseph LeDoux's *Anxious;* both the LeDoux and Bourke books were helpful in shaping my thinking about the different strands of fear, and the differences between fear and anxiety. The discussion of the primary emotions comes from Antonio Damasio's *Looking for Spinoza.*

Further down, I pulled Harold Kushner's count of the biblical commandments to "fear not" from his book *Conquering Fear.* The brief history of phobias and their treatments over the centuries comes from Helen Saul's *Phobias,* a book that offered a helpful overview of the main schools of thought and areas of research over the years. Ivan Pavlov's work was discussed in several of my source works, including Saul, Bourke, LeDoux, and Bessel van der Kolk's *The Body Keeps the Score*. (The specific details about the dogs in the flood come from Van der Kolk's book, as well as from Oliver Sacks's *The River of Consciousness*.) Likewise, John Watson's seminal study of Little Albert was widely discussed by my sources—I built the description of the experiment primarily from his own account, as well as from Joanna Bourke's summary. My description of Freud's study of Little Hans is pulled primarily from Saul and Bourke; I found the details of Freud's early work in neurology in Joseph LeDoux's *Synaptic Self*.

My brief summary of the workings of the brain is built largely on a much more comprehensive and precise description in John E. Dowling's *Understanding the Brain*. The estimate of the distance spanned by a single adult human's axons comes from Antonio Damasio's *Descartes' Error*. Joseph LeDoux's *The Emotional Brain* was also helpful in my effort to briefly describe the workings of some key brain structures. The description of the fear-reaction process is built largely from Damasio's *Looking for Spinoza;* discussion of the commonsense school of thought is from LeDoux's *Anxious*. I first came across the William James quotation in Damasio's *Descartes' Error;* the description of the Parkinson's patient is from *Looking for Spinoza*. Joseph LeDoux pointed out the Greek origins of the word "anxiety" in *Anxious*.

The discussion of the nature of dreams is built almost entirely from Alice Robb's *Why We Dream* and from Shelley Adler's original 1991 journal article on the SUNDS phenomenon.

Chapter Four

The middle section of chapter four is based on my interview with Lilianne Mujica-Parodi, as well as two journal articles resulting from her research: "Chemosensory cues to conspecific emotional stress activate amygdala in humans," from *PLoS ONE*, and "Second-hand stress:

Inhalation of stress sweat enhances neural response to neutral faces," in *Social Cognitive and Affective Neuroscience*.

Chapter Five

The statistics on the prevalence of acrophobia come from "Deconstructing Acrophobia: Physiological and Psychological Precursors to Developing a Fear of Heights," in *Depression and Anxiety*, and "Fear of heights freezes gaze to the horizon" in the *Journal of Vestibular Research*. My discussion of the potential evolutionary, genetic, or personality-based origins of phobias is drawn primarily from Helen Saul's *Phobias*, as is the description of the New Zealand–based study of children who experienced significant falls.

The specific study discussed in relation to my own fear of heights is "Fear of heights freezes gaze to the horizon," in the *Journal of Vestibular Research*, but my thinking here was also informed by "Deconstructing Acrophobia" and by "Fear of heights: cognitive performance and postural control," in *European Archives of Psychiatry and Clinical Neuroscience* (2009).

My description of Mary Cover Jones's work comes directly from her own account of the Little Peter study. Joanna Bourke's *Fear: A Cultural History* shaped my description of Joseph Wolpe's work; Bourke also provided the overall narrative of the evolution of treatments in this period. Many of the details about the heyday of lobotomies, including the quotation from Walter Freeman, come from Luke Dittrich's *Patient H. M.*, which I would highly recommend to anyone interested in the science of memory or the ugly history of medical treatments for mental illness. Stanley Law's description of electroshock therapy came from Joanna Bourke. Bessel van der Kolk's *The Body Keeps the Score* picks up the narrative of treatment where Bourke leaves off—with the rise of pharmaceuticals.

Edna Foa described her work to me in a brief telephone interview; I've also leaned on her explanations of concepts like extinction here.

Chapter Six

Maggie O'Farrell's *I Am, I Am, I Am* was helpful in shaping my thoughts about the meaning of near-death experiences; beyond that, it's just a beautiful book.

My brief overview of the treatment of PTSD through the twentieth century comes largely from *The Body Keeps the Score*, as well as from my conversation with Edna Foa. My description of the origins of EMDR is drawn from Francine Shapiro's *EMDR: The Breakthrough Therapy for Overcoming Anxiety, Stress, and Trauma*. Shapiro's original 1989 study is "Efficacy of the Eye Movement Desensitization Procedure in the Treatment of Traumatic Memories," published in both the *Journal of Traumatic Stress* and the *Journal of Behavior Therapy and Experimental Psychiatry*. The critical commentary that followed in 1993 was published under the title "Theoretical, Clinical, Research, and Ethical Constraints of the Eye Movement Desensitization Reprocessing Technique," by Julian Metter and Larry K. Michelson. Examples of the growing body of research that found some evidence for EMDR's efficacy can be found in "A meta-analysis of the contribution of eye movements in processing emotional memories," published by the *Journal of Behavior Therapy and Experimental Psychiatry* in 2013.

Anyone especially interested in the "resourcing" aspect of the version of EMDR that I underwent should check out Laurel Parnell's *Tapping In*.

The quotation from the survivor of the Oklahoma City bombing comes from Francine Shapiro's book. Bessel van der Kolk's comments about Pavlov's dogs and "inescapable shock" come from *The Body Keeps the Score*.

Chapter Seven

The story of the musophobic woman that opens the chapter came from the boxed text labeled "A reconsolidation-based treatment of a decade-old fear memory" in "Breaking boundaries: optimizing reconsolidation-based interventions for strong and old memories," a

2017 article by Merel Kindt and James W. B. Elsey published in *Learning & Memory*.

Karim Nader's research on memory reconsolidation was published by *Nature* in 2000 under the title "Fear memories require protein synthesis in the amygdala for reconsolidation after retrieval." The story of how Kindt applied the results of Nader's research in her own work comes from my initial phone interview with her, as well as from two key papers she and her colleagues published (2009's "Beyond extinction . . ." and 2015's "An abrupt transformation . . ."). The rest of the chapter relies on that initial interview and a second in-person interview, informal discussions with Kindt and Maartje Kroese, and my own experiences of applying for treatment and then being treated in Amsterdam.

Chapter Eight

My description of Alex Honnold's climb up Half Dome is drawn from his memoir, *Alone on the Wall*, as are his reflections on skydiving and his relationship to fear more generally. The details of his turn in an fMRI machine come from J. B. MacKinnon's *Nautilus* story, "The Strange Brain of the World's Greatest Solo Climber."

The details of Patient S. M.'s life and condition come primarily from "A Tale of Survival from the World of Patient S. M.," the first chapter of *Living Without an Amygdala*. The neuroscientist Ralph Adolphs made the comment about "creepily specific lesions" during our phone interview, while explaining Urbach-Wiethe to me more broadly; the statistics about the number of times studies of S. M. have been cited come from a table included in "A Tale of Survival." The Oliver Sacks line about deficits comes from *The Man Who Mistook His Wife for a Hat*.

The multiyear study of Patient S. M. that launched in 2003 was published in *Current Biology* by Justin Feinstein, Ralph Adolphs, Antonio Damasio, and Daniel Tranel in 2011 under the title "The human amygdala and the induction and experience of fear"—I've used both the study itself and the more expansive descriptions in "A Tale of Survival" to reconstruct the research process here. The same goes for Justin Feinstein's carbon dioxide study—I leaned on both the book chapter and the 2013 *Nature* article "Fear and panic in humans with bilateral amygdala damage," as well as on my phone interview with Justin Feinstein.

The discussion of S. M.'s protective maternal instincts, and the quoted dialogue with a researcher, come from "A Tale of Survival"—the dialogue appears in a boxed text separate from the rest of the chapter.

Chapter Nine

Gavin de Becker's *The Gift of Fear* has its limitations, but reading it offered me a new way of seeing some of the more harrowing experiences in my life.

My interview with Dr. Mujica-Parodi helped me to position her published research in relation to real-life experiences like the ones de Becker describes (although I didn't mention *The Gift of Fear* to her specifically when we spoke). The section of this chapter that follows up on her skydiving studies is based on our phone conversation as well as on her 2014 article in *NeuroImage*, "The fine line between 'brave' and 'reckless': Amygdala reactivity and regulation predict recognition of risk." I also appreciated Justin Feinstein's willingness to talk about the role of fear in our lives more broadly. The pairing of real-life experiences with research in this chapter felt especially tricky to me; the conclusions I've drawn about when and how to trust our fear are entirely my own.

The story about my friend's encounter on the bike path is shared here with permission.

Epilogue

The Michel de Montaigne line comes from Joseph LeDoux's *Anxious.*

acknowledgments

A version of the material in chapter five was originally published by *Esquire* under the title "Exposure Therapy and the Fine Art of Scaring the Shit Out of Yourself On Purpose." Thank you to my editor on that story, Megan Greenwell, whose enthusiasm helped give me the confidence that this could be a book. Thanks also to Tim Folger and Sam Kean, who selected the story for *The Best American Science and Nature Writing 2018*—another confidence boost.

Thank you to Lilianne Mujica-Parodi, Edna Foa, Justin Feinstein, and Ralph Adolphs for taking the time to talk through their fascinating research with me. Thank you as well to Merel Kindt and Maartje Kroese at the Kindt Clinics in Amsterdam not only for their important work, but for their time and their kindness. My visit to the clinic was genuinely life-altering. So, too, was my experience with EMDR; thank you to Mark Kelly and to Svenja. Sadly, Francine Shapiro died while I was working on this book. I regret that I didn't get the chance to talk to her about her invention.

I had the assistance of a small army of brilliant colleagues and friends in writing and editing the book. Kate Harris and Kate

Neville loaned me their off-grid cabin for two separate stints at critical moments: I wrote much of the first third of the book, and a year later completed some near-final revisions, in their little piece of paradise. Doug Mack's assistance with my proposal draft was invaluable. Adam Roy, Simone Gorrindo, Ferris Jabr, Frank Bures, Krista Langlois, Kate Siber, Elon Green, Sarah Gilman, and Cally Carswell all pitched in to read drafts of chapters; Lauren Markham and Brooke Jarvis read large chunks of the manuscript; and after also reading some earlier portions, Katherine Laidlaw heroically went through the whole thing at the eleventh hour. Katherine and Brooke bore the brunt of my book-related fears through this entire process, and I am so grateful for their support and friendship.

Jane C. Hu fact-checked the science-focused portions of the book, and her keen eye brought me some much-needed peace of mind. Any outstanding errors are my own.

Jennifer Weltz, my literary agent, was on board with the idea for this book from day one, and has been a fierce advocate on my behalf throughout the process of bringing it into the world. Thank you to her and to everyone else who pitched in at JVNLA.

My editors, "the Nicks," Nicholas Garrison at Penguin Canada and Nicholas Cizek at The Experiment, provided kind words, careful reading, and thoughtful feedback. The book is much stronger thanks to their attention.

My aunts Shelagh and Rosemary helped fill the holes in my knowledge of our family history, never hesitating to answer often painful questions about the loss of their parents. My dad, Doug Holland, also gamely answered any questions I had about his role in the story.

Thank you to everyone at Hospice Yukon. Thank you to Barb Lankamp-Kochis.

Many of my friends in Whitehorse have been a part of this story since I first began to admit my fear of heights. Thank you to Joel MacFabe and Nicolas Filteau for helping me down that frozen creek, to Ashley Joannou for encouraging my plan to skydive, and to Lindsay Agar and Kealy Sweet for taking me rock climbing. Maura Forrest not only became my primary climbing partner during my exposure therapy efforts, she also drove me to Alaska and back when I was too scared to get behind the wheel.

Ryan Agar and Carrie McClelland have been teaching me to face my fears and supporting my efforts in the wilderness for the better part of a decade now. It's no coincidence that they turn up at key moments in the book. Thank you to them, and to everyone else who has ever talked me back to my feet when I was—literally or metaphorically—curled up in terror on the ground.

about the author

Eva Holland is a correspondent for *Outside* magazine, and a former editor at *Up Here*, the magazine of Canada's far north. Her work has also appeared in *Esquire*, *Wired*, *Bloomberg*, *Pacific Standard*, *AFAR*, *Smithsonian*, *Grantland*, *Seattle Met*, *National Geographic News*, and many other outlets. Her work has been nominated for a Canadian National Magazine Award, anthologized in *The Best American Science and Nature Writing*, *The Best Women's Travel Writing*, and *Best Canadian Sports Writing*, and listed among the notable selections in multiple editions of *The Best American Essays*, *The Best American Sports Writing*, and *The Best American Travel Writing*. She lives in Canada's Yukon Territory.